British Library Cataloguing in
Publication Data

Glenn, John Albert
    Children Learn to Measure
    (The Harper Education Series)
    1. Mensuration – study and teaching
        (Elementary)
    i Title
    372.7'2   QA465   047741

    ISBN 0 06 318155 X ✓   374.51
        0 06 318156 8 Pbk
                              136040 gLE

Typeset by Inforum Ltd, Portsmouth
Printed and bound by A. Wheaton &
Co Ltd, Exeter

CHILDREN LEARN TO MEASURE

Foundation Activities in the Classroom
(5 - 11)

A Handbook for Teachers

THE MATHEMATICAL EDUCATION TRUST

CHILDREN LEARN TO MEASURE

Foundation Activities in the Classroom (5 - 11)

A Handbook for Teachers

**Edited by
J.A. Glenn**

Illustrated by David Parkins

Harper & Row, Publishers
London

Cambridge   San Francisco
Hagerstown   Mexico City
Philadelphia   Sao Paulo
New York   Sydney

Contents

Preface

This book, the fourth volume issued by what is now the Mathematical Education Trust, follows the general plan of the earlier *Children Learning Geometry* (Harper & Row 1979), adapted as necessary for a different topic. It covers a primary course in measure for children in the age range five to eleven, and is compiled as a working manual of activities for the teacher in the classroom. Like its predecessor, it relies on the teacher's professional skill to present each activity in the form appropriate for each group of children.

Although the book is written for the experienced practising teacher whose main interest may not be mathematics, there is a general introduction that covers a wide range of preliminary discussions which we hope will help those undertaking in-service training. For those in initial training the book should help prepare lessons or work schemes for primary pupils. It should also provide material for those planning or running in-service courses.

A major contribution to the book has been made by Leone Burton of the Polytechnic of the South Bank, whose work suggested the general plan of the activities. Other useful suggestions and activities have come from

Geoffrey Austwick   Gwen Renton
A.M. Davies   Peter Scopes
Ronald Golding   Peter Smith
Jo Lister   Derek Woodrow

The book has been compiled and edited by J.A. Glenn of the former Kesteven College of Education, now of the British Campus of the University of Evansville.

The Trust is indebted to the Nuffield Mathematics Continuation Fund for a grant meeting the expenses of preparing the manuscript and to Sheila Harris for typing the drafts.

The editor and publishers are grateful to readers of the first draft who made many useful criticisms and suggestions: Hilary Shuard, Richard Harvey, Hilda Doran, Peter Dean, and Shamus Dunn.

Part 1

The background to measure

Mathematics and measure

Since mathematics began to be taught in schools the word has tended to carry two meanings – the one it has for the mathematician and the 'man in the street' meaning of proficiency in arithmetic. Although there is a serious shortage of people with developed mathematical skills, it is arithmetical deficiency that is of immediate social importance.

Arithmetic in its turn has its two aspects. There is the study of number and the operations involving it, and there is the applied arithmetic of the older textbooks, based almost entirely on manipulating the once complicated units of money, length, area, capacity, and so on. This work in measures often played the part of cuckoo in the nest, displacing mathematical studies proper before they had a chance to develop in the classrooms. Many schools, having laid down the Four Rules, devoted their lesson time to long calculations in the former imperial units. Pendlebury's famous *Shilling Arithmetic* of 1899 fills some 150 of its 200 pages with such examples.

Although measure as such is used in science, technology, and trade rather than in pure mathematics, there is no suggestion that the mathematics teacher should disclaim responsibility for it. Skill in working with the commonly occurring measures on whose numerical expression the Four Rules operate is *de facto* in the primary school mathematics schemes and has long been accepted without query. For all that, our work with children will be less effective if we do not realize that when teaching measure we are operating at two levels.

However simple the numerical examples, they already belong either to abstract mathematics or, in the widest sense of this word, to physics. Consider

$$2 + 4 = 6$$
$$2 \text{ litres} + 4 \text{ litres} = 6 \text{ litres}$$

If we do not stop to think, the second sum seems to be merely arithmetic in action. But the litre, as a measure in actual use, can only be identified by a process that compares a given quantity with a standard quantity. Putting this process into operation is not mathematics, which only takes over when we want to operate on the two numbers 2 and 4 which are linked to the quantities concerned.

We speak of the conservation of volume, and some psychologists are inclined to treat this as logical. It is, in fact, empirical. Two litres of water poured into four litres may well give us six, but what happens if we use the water to dilute four litres of, say, alcohol? The answer cannot be discovered by arithmetic alone.

Much the same is true, in practice, of other measures. A piece of string shortens considerably if wet, doors rattle in summer and stick in winter, clothes stretch and shrink to a point at which they cease to fit, bottles of scent evaporate even when stoppered. We do not, of course, bother the children with these quibbles. The conservation laws as we normally introduce them into the primary classroom are of the utmost importance – an essential step in setting mathematics to work. We want the children themselves to use them with confidence, not speculate about their logical status. The adult user of measures soon comes to grips with their practical limitations.

But measure has an importance for mathematics that our discussion so far has ignored. Many basic mathematical concepts come in the first place from working with number in the physical world. The mathematician, and certainly the child who is beginning to learn mathematics, actually needs measure to help construct the world of numerical and spatial relationships from which the study arises. We do not begin to form clear concepts of length, area, angle, and so on until we begin to measure them. Looking back on our knowledge when we are older, these concepts appear almost intuitive. Plato felt able to argue that they are born with us, vague memories of an earlier and more perfect existence. Today we are aware of how slowly children's ideas of number and space develop.

Measure is the meeting ground between mathematics and its applications. It is measure that allows physics to become quantitative and hence open to mathematical treatment. It is measure that provides mathematics with its most far-reaching starting points. The concept of a fraction almost certainly came from the failure of a fixed unit to measure all lengths or masses, and this is still a valuable discovery for children to make as a result of carefully planned activity.

Measure is also, as old Pendlebury saw, an obvious field for applying numerical skills. It is unfortunate that many of his exercises seem to have been thought out in his study and had only a marginal relevance to the industry, trades, and commerce even of his own day. At the best, they only call for calculation with given measures: there is not one exercise in the whole book that asks the pupil to measure something. We can now, at least in our primary schools, do much better by letting our pupils obtain the measurements on which to put their number skills into action. Mathematical education can find only profit in its adoption of measure as a classroom topic.

Pendlebury and other well-known writers of nineteenth-century arithmetic books give no hint of the fundamental importance of measure to abstract mathematics, and indeed the mathematicians themselves did not admit, or perhaps did not then see, how deeply indebted they were to the utilitarian development of scales and measuring rods. Measure, in fact, enables us to count what is not at first seen to be countable. We can count eggs and apples because they are discrete and individual objects; but how could one 'count' and hence attach a number to such a thing as a length of rope or a sheet of paper? The whole of modern analytical mathematics springs from the ancient insight that one should be able to use one quantity as a unit to quantify another; that is, to use it as a measure.

Measurement, of course, is useful and was only developed because of its use. But to teach measure to children on the sole grounds that it will get them a job is misguided. We want our work in measure to be relevant to its actual uses, yes, but we are not training primary school-children to become surveyors or cabinet makers. The technical uses of measure are not the province of the schools; but the interaction be-

tween mathematics and measure certainly is. As such, it is a necessary component of mathematical education.

At the primary level, indeed, one can argue that the main reason for getting children to measure is to develop their understanding of the number system and its notation. Measures in use provide tangible models of the grouping of units into bigger units and of the subdivision of units to enable us to measure more finely. Now that our measures use units related in tens, measurement shows the decimal notation of hundreds, tens, units, tenths, hundredths . . . actually at work.

At one time people found it hard to accept the metric system because they 'did not understand decimals'. We can now see that the primary schoolchild has, in the subdivision of the metre into decimetres, centimetres, and millimetres, a working model of decimal fractions. Schools are only just beginning to profit by the general introduction of metric units.

In home and family

In our eagerness to justify mathematical studies to practically minded people we tend to exaggerate the daily uses to which measure is actually put in the home, and certainly to exaggerate, in constructing 'examples' for children to calculate, the precision to which one normally works.

Consider, as an example, the making up of curtains. The material is only obtainable in fixed widths and is sold in lengths only to the nearest $\frac{1}{4}$ m. The depth of the window is measured, probably to the nearest centimetre if a metric tape is used; but when this measure is transferred to the material a generous allowance is left, almost always by eye, for a deep hem. This hem is usually adjusted after the top hem is made and tapes are fitted so that the curtain hangs at the right level, and is more often than not pinned up by eye. The hem is made deep so that it can be let down if the curtain shrinks on washing, by an amount almost impossible to decide in advance. There is a lot of skill and judgement required to do the job properly; but it is all skill acquired by practice and experience: the initial task of measuring the window is relatively trivial. The depth and width of the window are necessary data with which to begin, but only manual skill and acquired judgement carry the task to a close. One remembers the suit of clothes 'very ill made' provided for Gulliver by the tailor of Laputa after taking that critical traveller's measurements with scientific precision.

Any home in which cooking, dressmaking, paper hanging, shelf construction and so on is done by members of the family can provide endless examples, but in each of them the skill called for by the use of tape measures or kitchen scales is small compared with the overall expertise required by the activity itself. Children whose parents have cars will certainly hear journeys discussed in terms of mileage and petrol consumption; but here the use of measure is purely descriptive. In these days we all think of travel in terms of time rather than distance: the brochures for holiday resorts abroad tell us how long the journey takes, not how far we have to go and at what speed.

Most children at school are likely to do more measurement, and by the end of their primary school days to be doing more accurate measurement, than they will need for running their homes when they grow up. Between the ages of five and eleven the skills they acquire in the primary school are certainly adequate for domestic needs, although whether they will still have them in their twenties or thirties when wanted is a different question. As for calculation using constructed examples, it is general number skills, particularly with decimals, that will carry them through most adult requirements. It is interesting to skim now through Pendlebury's 150 pages of examples trying to pick out those that are now used in daily life in the form given, either in home, office, or industry.

We conclude that we do not teach measure merely because it is going to be useful in the home. It is useful, of course, but only as a modest skill easily picked up by anyone who really needs it. The fundamental reason for teaching measure is an educational one, and it is only incidentally a preparation for householding.

Industry and science

We have already discussed the computational needs of entrants to employment in our earlier report, *The Third R: Towards a Numerate Society* (Harper & Row, 1978 – see back cover). The required skills in measure would seem, except in so far as they are most general, to be an obvious topic for initial training or observation periods on entry to a job involving them. Some school leavers, who happen to have chosen school-based skills such as woodwork, dressmaking, or technical drawing, perhaps with a related occupation in mind, will probably have as much expertise with measurement as an entrant on apprenticeship or training will need. The use of an apparatus designed to do a special job, such as the dial measure now found in many fabric shops for taking lengths of material accurately from the roll, can easily be picked up by a retail assistant after a few demonstrations and practice runs, and is clearly a matter for the employer or training officer.

What is needed is a general familiarity with measures of length, mass, capacity, and time, using the simplest apparatus. Not enough trades or occupations use measurement of area and solid volume to justify the teaching of these topics to *all* children on the sole grounds of occupational need. The educational reasons for a thorough treatment of area and volume in schools are very strong indeed, and apply to children in primary schools still many years from careers or job seeking. We need a general familiarity with the concept of measurement that will later be called for by industry. All specific skills can either be brought into industry by school leavers who have taken practical courses at secondary level, or can be dealt with in the context of the actual job by the industrial training officer or employer.

Our earlier comment on the comparative triviality of domestic measurement skills seems to apply even more strongly in industry or trade. Someone who cannot measure and mark out a length accurately along a plank is unlikely to be able to use a handsaw to cut accurately and squarely on the correct side of the line, even if someone else did the measurement. Anyone who has tried to fit a shelf into an alcove will appreciate the point. The example is of the sort often quoted, although it is obviously taking us back to domestic rather than modern industrial demand.

Industry, the large, loose productive organization on which our well-being depends, now has little room for homespun craftsmanship. Production involves measure at very high levels of precision in dimensions, tolerances, physical properties, and quantitative control. It should be the task of industry, even small-scale industry engaged in specialized production, to see that its entrants are trained to cope with what is specifically needed, but it remains the responsibility of the schools to help children learn the general skills and acquire the fundamental concepts that are brought into play.

The needs of science are rather different. Industrial science is clearly a matter for specialist skills and knowledge, but the exact sciences have long been part of the school programme. During this programme pupils learn to make measurement to a more than domestic degree of accuracy, and go far beyond primary activity. They use the fundamental measures to make derived measures such as density, acceleration, and potential difference, selecting at school level from the whole range of quantitative results that provides science with its raw data. Such derived measures are conceptually difficult and highly specific to the various branches of science.

These measures and the techniques for making them at an elementary level are clearly the province of science teaching; but there is no doubt that the fundamental measures are best taught along with number work. We shall return to this in a later chapter on children using measure in primary science (page 114). What we should accept without question is that by the time pupils begin secondary science courses they should have a firm conceptual grasp of the fundamental measures of length, mass, and time, a good practical knowledge of capacity, and a familiarity with temperature. They should have these not merely to prepare them for their courses, but as part of their general development. Their practical measurement skills would no doubt need to improve in accuracy, but this is one of the objects of the laboratory work they will do. Given this (and it does not follow that *all* children can reach it) any difficulty with the quantitative sciences suggests that the pupils are really failing at activities of a higher level of conceptual or practical difficulty.

A note on units

It is difficult for Britain to introduce reforms without controversy and delay, probably because the course of history shielded us, almost throughout the development of modern industrial societies, from the need to adjust ourselves to conventions that were elsewhere becoming slowly international. An example is the change from imperial to metric units. A proposal to go metric was defeated in Parliament by five votes in 1871, although a little later the metre became legally recognized if not adopted. We changed to a decimal currency at the last practicable moment, and have avoided the international convention of a right-hand drive for road vehicles for so long that it is now almost impossible to change. Officially we can now use the metric measures as described in the Système International (or SI), although some trades continue to use the narrow range of old imperial units – the word itself serves to date them – that happens to apply to their work. Very few younger people now have any knowledge of the imperial system beyond this limit of personal contact. The present legal definition of the units that remain in use is in terms of the metre and kilogram, as it is in the United States.

The activities in this book, apart from the early reference to cupsful, paces, and so on, will use only SI units. By the time even the youngest pupils we are dealing with go out into the world, a few of the older units will no doubt remain in customary use, but already none of them is used in computation. Our petrol pumps still read gallons but have long adopted a decimal subdivision, and bulk delivery to garages is in litres. The supermarket weighing machines still show pounds and ounces but work to decimals of a pound when they are designed to show the cost as well as the weight of a purchase.

We no longer need to argue a case. Some of the SI units, like milligram and microsecond, are not within the range of magnitudes that can be handled by children, and fundamental units such as the ampere and the mole are well beyond the scope of primary studies. We shall use only the units listed below, and it is worth noting that these, which could be written out on a postcard, not only meet the needs of the primary child and all ordinary domestic measurement but, indeed, most of the general requirements of commerce and industry. There are seven fundamental units, but we are concerned with only three. These three units are very accurately standardized, and act as reference measures for all the practical units we may wish to use.

We give them in this list with four 'customary' and two derived measures. No others need be taught at primary level.

Of the units given, the degree Celsius is customarily used in place of the fourth fundamental unit, the kelvin (K), a measure of temperature that need not concern us. The degree Celsius is often called the degree centigrade in the UK. Since the centigrade is also a unit of angle once used widely throughout the rest of Europe, we are adopting the more usual name, given in honour of its innovator Celsius. Fortunately this name also begins with C. Of the other measures, 1000 litres are equi-

Quantity	Unit	Symbol
length	metre	m
mass	kilogram	kg
time	second	s
fluid capacity	litre	l
angle	degree	o
temperature	degree Celsius	°C
land measure	hectare	ha
area	metre squared	m²
volume	metre cubed	m³

valent to a metre cubed and one hectare is 10 000 metre squared. In ordinary speech metre squared and metre cubed are usually called 'square metres' or 'cubic metres'. This is probably the best form for everyday use.

The SI units use a wide range of multiples and submultiples of the fundamental units, but only a few of them are needed for the general purposes of society. These are the multiples and submultiples we shall introduce through the ensuing activities. We can take it that nobody is likely to use any measures other than those collected in the following list unless he is in an occupation with specialized requirements. The list may be compared with the long lists of trade and commercial measures given on the backs of exercise books fifty years ago, or the six printed pages of tables in Barnard Smith's *Arithmetic*, as revised in 1865. Note the alternative names for the units of area and volume; these are not 'official' but are commonly used and generally understood.

The megagram is 1000 kg and has long been known as the 'tonne', or 'metric tonne' in English-speaking countries. This word will continue to be used in place of the more formal name. The prefix 'mega' is often

Length	Area
kilometre (km)	metre squared (m²) (or square metre)
metre (m)	centimentre squared (cm²) (or square centimetre)
decimetre (dm)	millimetre squared (m²) (or square millimetre)
centimetre (cm)	
millimetre (mm)	

Mass ('weight')	Volume
megragram (Mg)	metre cubed (m³) (or cubic metre)
kilogram (kg)	
gram (g)	centimetre cubed (cm³) (or cubic centimetre)
milligram (mg)	millimetre cubed (mm³) (or cubic millimetre)

Fluid Capacity	Time
litre (*l*)	all common units (seconds, hours,
centilitre (c*l*)	years, etc.)
millilitre (m*l*)	

Angle	Temperature
angular degree (°)	degree Celsius (°C)

seen in newspaper reports giving the output of power stations in megawatts. Note that the symbol (although not the name) for the multiple unit begins with a capital letter M, for the submultiple unit with a small letter. Gramme and gram, with the derivatives of these words, are both acceptable spellings: we shall use the simpler (and originally American) form. Originally the SI system was very formal, and only specified submultiples in terms of a thousand, as metres/millimetres. It was soon seen that submultiples such as decimetre, centilitre, and so on were of great practical use, and, although scientific and technical writing sticks to the official multiples and submultiples, the use of the other convenient measures is now encouraged, particularly in primary classrooms.

A note on symbols

The names and symbols for the units are internationally agreed, and the symbols are used by all countries, whether or not their languages are normally written in the Roman alphabet. All industrially important countries also use the European form of the Arabic numerals, parallel with their own if these are different. It follows that 15 kg or 3m² is good Russian and good Japanese as well as good English or German. Children are usually interested to hear this, given that they are old enough to appreciate differences and similarities of language and culture. Since it usually follows a number, the symbol for litre has caused difficulty by confusion with the digit 'one'. Five litres can be read as fifty-one. For this reason a script or ilatic l is often used.

Because they are symbols and not abbreviations, they do not take a full stop or need a plural form, and it is wrong to write 15 cms for 15 cm as if it were an English word. Since the names of the units are also international, and since few of the world's languages show grammatical number in the same way, it has also been agreed that the names should not inflect to form plurals. Metre, then, should behave like the English word 'sheep'. Teachers might care to follow this usage in writing or speech, but we think the distinction is rather pedantic for pupils in school (even if many greengrocers and others often refer to 'three pound of potatoes'). Children and others will say litres and it is better to accept this. In discussion among ourselves it is difficult not to add the 's', and we use it in talking of units as on page 59). If one is later concerned with formal science, the change to an international convention should not be difficult. We should assimilate the SI system to our own language.

Measure in the primary school

It is only as a foundation study that primary work in measure is relevant to the practical needs of an industrial society. It is nevertheless relevant to the whole of our experience: this is a different kind of relevance.

Knowledge and understanding of distances, both astronomical and terrestrial, have become part of our daily lives, and indeed many of us now cover greater distances during a holiday than some of our ancestors would have actually travelled in a lifetime, although of course they walked far more from day to day (The ability to compare, measure, and estimate measures gives us an insight into our environment and is one more pathway leading to an informed appreciation of it. An awareness of the approximate nature of all physical measurement is of the greatest importance in this context, and may well give meaning to the use of the word 'informed' in the last sentence.

Measurement is also an activity that can be readily organized in schools and planned at any level of ability or development. As a practical activity, it is usually enjoyed by many children if properly presented to them. We hope to show that measure can inspire a programme of classroom activities that most pupils can enjoy.

There is also another broadly cultural aspect. The historical development of standard units of measure, set in the context of the development of modern society, is indeed a topic having intrinsic interest. It is not easy to persuade either the makers of secondary history syllabuses to modify their accounts of the Norman Conquest by including a discussion of the measures used in The Domesday Book, or to make those responsible for examination mathematics take a look at the history of the metre. As a result, the development of measurement as a commentary on social development in general is rarely discussed among older pupils.

In the primary school, not yet in the strait-jacket of externally imposed subject divisions, much could be done with the history of measurement, and this could well absorb some of the time once given to interminable calculations in imperial units. A full account of the history of measurement is one of those deep but narrow specialities that make such fascinating reading for the few who are stirred by them. For the teacher it can be a valuable source to explore for inspiration in looking for effective approaches to the work of a class (see Bibliography).

It will be useful to summarize here the goals in teaching measure as part of mathematical education, itself a term now much more widely interpreted than the more traditional (and still important) view of mathematics as expressing relationships and solving problems. The reader may or may not agree that these are given in order of educational importance; but in the work of the classroom one would like all of them to be kept in mind by the teacher.

1   To help understand and appreciate our immediate environment and its wider context by allowing us to think of it in quantitative terms where these are appropriate.

2   To develop an ability to estimate and measure the quantities that arise in daily living. Usually estimation and judgement are more important than precise measure.

3   To provide a background of agreed general skills and concepts that will, as a foundation, meet the specific needs of any occupation or training programme. It is to be understood that specific techniques of measurement called for will be developed in the context of the activities concerned.

4   To provide intellectual stimulation and enjoyment.

5   To show the story of measurement and the development of standard units as a typical theme illustrating social history.

6   To understand the properties of measure, particularly of length and area, in such a way that they form a conceptual basis for further studies in mathematics at a specialist level.

This list is given to help analysis and discussion when teachers are deciding on topics and items in a work scheme. It is not intended to separate out all the distinct possible factors that make up a programme for teaching measure, but only to draw attention to some of the ways in which we can approach it. Not all of the headings need apply to each and every topic being considered. What is most important, perhaps, is the loose distinction to be maintained between the third goal above and its expansion into technology. With one of those enthusiastic groups whose emergence in the classroom is the delight of the primary teacher, one could well investigate chosen measures more deeply than ordinary domestic use and background knowledge require, but there should be no pressure to do this.

Standard units

A standard unit is one that, like the British imperial or ancient Roman units, is determined by law or general agreement and can be used for domestic or international trade to provide a uniform system of measure. The points at which such standard units of measure should be introduced into the syllabus need discussion, particularly as some children will have seen measure in use at home in simple imperial forms: pounds, ounces, gallons, pints, yards, feet, inches. (These were usually with a wide range of conventions superimposed. Thus, curtain material would have been bought by the yard and fractions of a yard, but measured individually in inches only.) Such children, perhaps at the bidding of parents, may well ask 'When are we going to use pounds and ounces?'

There are two problems here, what to do about standard units and also what to do about imperial measures, which for some time to come are likely to be the ones, if any, talked about at home.

We think that specimens of the imperial measures should be available to children from about 7+ upwards, but they should not be frequently used. We want the children to think of them as museum pieces. The children can handle them, use them for the sake of interest if they wish; but they will no longer work with them or have table learning sessions. Where a measure is still in nominal daily use, as the pint is for the standard milk bottle or the gallon for petrol, the child learns about these more adequately at home than at school.

We would suggest a permanent display in a school corridor or the hall, suitably labelled (much as bushels are today in folk museums), with other specimens for actual handling in the classroom.

The current modern practice is to get children to work with arbitrary measures such as spans or personal cubits, so that they can see the difficulties of making comparisons and recording permanent results. The children are told that early man used these, and the standard units of measure are then produced as the ingenious solution to the problem of using measure socially. We may need to reconsider the practice.

The very earliest recorded measures were at least regionally standard, so that the sequence is at the best hypothetical history on the lines of the Just So stories. Although it is science rather than mathematics that suffers most from this approach by the recapitulation of development, any historical setting in the early stages is suspect because young children have as yet no sense of a historical time scale. Dinosaurs and ancient Britons are more or less contemporary, and both not very far removed from great-great-grandfathers. It is, moreover, unlikely that young children would invent the standard unit for themselves. If Paul and Pam each pace the classroom and get different results, their probable response to the teacher who asks them how the difficulty can be cleared up is: 'You pace it for us.'

We want, nevertheless, to be able to separate the two concepts in 'standard unit', and the early work in arbitrary units is educationally valuable. For the purposes of these activities we shall use the sequence: arbitrary units, then standard units, postponing discussion of the advantages of standardization which can be introduced later in historical terms. Teachers who do not agree can make their own transition in using the activities.

The home background

Any schemes of work for children who are just beginning school have to face the difficulty of choosing starting points. A five-year-old with a lively and active home life is on the educational ladder before beginning school and sometimes finds it boring, while a less fortunate child brought up without helpful parents and well-chosen toys may find the demands of the reception class too much to grasp.

The infant teacher, fortunately, is well briefed in this matter, and acquainted with strategies for checking on the background of knowledge and experience that may at one extreme be missing or at the other well established. This background is certainly important for early work on measure, and needs to be fitted in when it is lacking. Many children can count objects from 1 to 10 without mistake before they start school and some can get as far as 20. Such children can be asked to pace across a room and count their steps: the results may be erratic but lead directly to the concept of a unit of measure and its use in comparing distances. A child who cannot count cannot use measure in any quantitative sense, but can begin to discuss objects using relations such as 'longer than', 'heavier than', 'the same size as', and so on.

Small children whose parents have been willing to endure their help in cooking or dressmaking may have seen measures in use, or may have used them in a haphazard 'make-believe' way. This will be true for the three measures of capacity, mass, and length, and probably for time intervals as well. Children like waiting for the kitchen pinger to go at the end of a cake-baking session, becoming visibly intent as the dial marking pointed out to them shows that the bell or buzzer is getting ready to sound. We note that their interest is only aroused when the set interval is nearly finished: this kind of observation tells us much about a child's subjective feelings for the passage of time and should affect the way the topic is approached in the classroom.

Because our approach to measure is through organized activities, often involving groups or pairs of children, the young child also needs qualities one would associate more with ordinary social development

and an ability to get on with teachers and classmates than with abstract mathematical concepts or even a personal vocabulary. Mathematics is a human activity like anything else done by pupils, but it has tended to be thought of as an essentially private structure of knowledge. We want children to work together to do mathematics. Without the freedom to move about and handle things, lacking the confidence to recover from mistakes or incidents such as breakage, the qualities we are after will be slow to emerge. The fundamental operation behind all measure, which makes the concept of measure possible in the first place, is the coordination of visual and kinaesthetic experiences, matching in our brain and our thinking what is felt in our muscles and seen with our eyes.

Because of their home background, many children are already better coordinated and are ready to start with more formally structured classroom activities than are others, who need to gain the experience earlier absorbed by their classmates. The range of work that needs to be planned is very wide, and has in part dictated the form taken by the activities that make up this book.

Sequence in learning

A glance at any early arithmetic book shows several measures being taught in a traditional sequence. Length is introduced first, then area and volume, and finally liquid capacity. In any infant class, however, children may be seen pouring liquids in play sessions, and such children can begin to formalize the concept of capacity long before they can appreciate volume as a count of unit cubes.

Children become aware of spacial relations long before they can think about them, and adults are often tempted to interpret a child's perceptions in adult terms. Clearly the floor enclosed within a playpen offers less opportunity for action than the whole floor of the room, and many of the most interesting objects are always out of reach; but it is rash to say that in this way the toddler becomes aware of area and distance.

It is safer, and more productive educationally, to say that the child's world is not analysed dimensionally at all: the concepts of length, area, volume are abstracted from it by the later activity of measure. Where there has been no suitable activity, the precise concept remains sterile. It may even be replaced, if called for practically, by an equivalent concept of a different kind. In The Domesday Book, land is measured in terms of the time taken to plough it, and the actual areas recorded may well have varied with the terrain: a very sensible arrangement in view of the use to be made of the concept.

The undoubted mathematical hierarchy of linear, square, and cubic measure was abstracted by geometry from a world of irregular solids. We would like children to be taken through some of the earlier stages prior to these abstractions. There is a distinction between the initial concepts and their final formal expression. Distance is not the same concept as length, room to move about is not the same as area, size or bulk is distinct from volume, and is not readily separated from the concept of mass.

We have here our key thesis: that children must learn to measure before they learn to calculate with measures. The traditional textbook merely used measure as an excuse for arithmetic, and many children left school quite unable to use simple instruments with understanding and competence.

The objects we handle have qualities that we learn to recognize and name using words whose meaning is relative to our day-by-day experiences. Things are big, tiny, soft, cold, long, hard, small, flat, near, distant, round, rough, smooth, tall. . . . Certain of these qualities can be quantified by the process of measurement: we can say not only that a man is tall, but how tall. Then we can *use* this measure if we want to buy him a coat. Number goes into action.

The activities in this book will be arranged by stages of mental development rather than in a hierarchy imposed by mathematics, and any one measure will be brought in or put aside as the work suggests. Words such as 'acre' or 'hide' do not suggest a hierarchy, but clearly 'metre squared' or 'millimetre cubed' not only suggest but demand sequential treatment. This sequence will emerge from the later activities. In the very early stages we measure whatever can be measured by such methods as the child's developing skill makes possible. Only later do we sort over and classify the measures, ready for their use in daily and domestic life and their extension to the technical needs of science and industry.

Part 2

Children learn to measure

A programme for measure

The activities that follow are presented in a similar way to those in our earlier *Children Learning Geometry*. That is, they are written for the teacher, and we rely on the teacher to interpret them, modifying the basic activity to suit the ages, abilities, and size of the groups of children who will use them. Only in this way can a practical activity be set down and used over a wide range of reading ages and differing classroom organization.

There is, however, an imposed sequence. In the activities with shape and configuration one could easily pick out different levels of approach, and some activities given early or late in the sequence could readily be transposed, adapting the work to another age group or to less or more developed pupils. For this reason the geometrical activities, although necessarily printed in a sequence, could be re-arranged at the discretion of the teacher. In this book the work is arranged in loosely defined stages.

There has been a tendency in education to define stages in learning which are occasionally thought or spoken of as an absolute classification. In so far as this protects children from approaches which are beyond their ability to grasp, the discussion of such stages is invaluable; but if it imposes standard patterns on learning schemes it can become as restrictive as mathematical sequencing. It does, however, make sense to think, for example, of a nonnumerical stage in which magnitudes are compared verbally, passing into numerical stages involving in turn whole number and then fractions.

We have decided on five of these stages, but are not greatly concerned with defining them. The five stages between them span the age range from five years to eleven. How teachers organize them for ability groups is a matter of school policy: the activities themselves do not dictate any procedure. We should prefer slower workers to spend their time working through four stages only, rather than struggling less effectively through all five, and at all stages we have tried to include a few activities that will stretch the abilities of the more skilful pupils.

The word 'activity' has been used in a very general sense. Many of them, as one would expect of classroom activities, are begun and brought to a conclusion, but others describe the sorts of things that children will be doing incidentally and are labelled and numbered as activities to make sure that nothing is omitted from the children's experience. Some topics, given as activities, are clearly intended to carry on over quite long periods, returned to again and again until familiar.

The vocabulary checks are an essential part of the structure. If, as Newton said at the beginning of his great *Principia*, 'the meaning of words is to be determined by their use', then we need to satisfy ourselves that children learn to use, with confidence and accuracy, the language with which we express the relationships formalized by measurement.

People often claim to understand a topic but deny being able to put their understanding into words. Whether this can be as true of a technical or academic exposition as it may well be true of an emotional or aesthetic reaction is not the issue; we are here concerned with the converse. If someone can explain something clearly and efficiently to another, can say exactly what is to be done and how, then we can hardly deny understanding.

One can usually find exceptions by seeking them out, and we intend the language criterion to be a working principle, not a general statement. Children of the travelling families living in caravans, and rarely attending school, no doubt soon become self-sufficient and skilled at the practicalities of life, yet make little use of formal codes of speech. Such groups are special cases with which books such as this cannot cope.

In *Children Learning Geometry* we have based evaluative procedures on the child's use of words, and the process has been extended for this volume on measure. We use the word evaluation rather than test because the object is not to test the children but to find out how effective the work done has been. What is being put to the test is the teacher and the method, as well as the ability of the children, and the result of such evaluation may show that somewhere along the line something has failed, calling for action before the next stage can be begun.

Perhaps we should say that the activities will become understood by use. Teachers may well have to work with them for a year before feeling confidence in their value. Because the activities are rather more strictly sequenced, at least between stages, than the geometrical material, the strategies for the classroom are not quite the same. We give our suggestions in numbered paragraphs as in the other book.

1   Read through the activities collected together for the stage appropriate to the children concerned and select one that is suitable for the circumstances of the lesson. For the early stages it is better if most of the children can work through most of the activities.

2   In *Children Learning Geometry* we suggested that the work could be spread over the age range as the teacher thought fit; but we recommend that work in measure should be done regularly each week. There is, however, no time limit to an activity, which may be extended or curtailed as convenient, as long as the evaluative targets are met.

3   Make sure that all the necessary apparatus and materials are available. It is a dubious teaching strategy to tell children not to bother about some section or other of the instructions given them because the material is not available. We make very little use of expensive purpose-built apparatus.

4   Use the evaluation checklists whenever the opportunity arises, with individuals, groups, or classes. The programme does not dictate how these are to be used, but we hope that none of the key words will be passed over. Although the lists are collected at the end of the stages, we hope that they will be used concurrently with the activities.

5   Consider very carefully any new words or phrases that the activity will need, and arrange for time to introduce these words, discussing their spelling and their use in context. The evaluation checks will show if the children can use them with understanding. Any new words, especially those specific to mathematics and not brought in from other topics, should be brought up in class several times over a few days, and thereafter kept in use as far as possible. Children forget new words they do not use as adults forget the vocabulary of a foreign language.

6   Try to anticipate difficulties that the children may find and questions that may be asked. Note particularly any difficulties that actually do occur. It is better if colleagues can discuss these together and agree on tactics for overcoming them.

7   It is not necessary to keep a record of each activity as each pupil completes it. Many of the activities are parallel and all of them represent skills or insights that need to be kept alive. The use of measure is very much a cumulative skill, with each stage dependent on what has gone before, and although each evaluative session tends to recapitulate earlier skills, it is worth recording the child's success as the work proceeds. We suggest a simple five-point scale of concepts, skills, and vocabulary, with the biggest group of children at C.

A – Firm and confident grasp of principles and skills.

B – Above average for the class.

C – Reasonable (i.e., as much as I could have done at that age).

D – Slow to grasp what is wanted.

E – Disappointing. Child may need remedial teaching.

This is a book on teaching, not assessment, and the object of the evaluation sessions is to determine the effectiveness of the methods used.

The reception class

The reception class is unique in education in that it can claim no prerequisites of knowledge or skill that some other teacher is supposed to have been responsible for. It is the first point at which we as teachers have any control over the environment offered to the children, the point at which deficiencies or abnormalities of home experience or speech first show up in terms of schooling.

Some children, of course, go to playgroups or nursery schools, but whatever their background the reception class has to take them as they come and begin their formal education. The work of the class cannot be readily divided into stages or topics: everything that is done is a general preparation, and it does not help much, except perhaps to ensure that nothing is forgotten, to speak of vocabulary work or manipulative activities.

As far as measure is concerned, the children will do many of the things listed under Stage One, but they will not be done in any formally organized way and they will be normally very repetitive. It is not enough to present an idea or a word today and, even if it seems to be understood and acted on, assume that it will still be properly known tomorrow.

The same notions, related for our purposes to the concepts of length or weight or capacity, will need to appear with many variants on many occasions, so that their meaning can be gradually deepened and extended. The Stage One activities collect and consolidate these notions.

Consider, as an example, the notion of length and the word 'long', as it would be used informally and linked up to all kinds of activities. The toy giraffe has a long neck, the doll a long nightie. The drinking straw is long, so is the telephone wire crossing the playground, and the line of boys and girls, and Leone's hair.

We can make a long line of the sitting blocks and go for a long walk in and out of them. We can go for another long walk back again, fore-shadowing in this the symmetrical aspect of measuring distances. We can unroll a ball of thick string, and perhaps make patterns by winding it around pegs; we can wind it up again. We can stretch elastic to make it longer (perhaps even noticing that it gets thinner). We can watch a bean or hyacinth shoot grow day by day, and see how much taller one child is than another when they stand side by side.

We can see that all new crayons are the same length whatever their colour, whereas the old ones are not; that Marie has shorter hair than Peter. We can also see that a very long ribbon laid out on the ground would prove to be a very short walk, that quite a small tree is bigger than a tall teacher, that a deep puddle is not the same thing as a deep pond.

The teacher can structure or make incidental use of similar situations for all dimensional words: high, wide, deep, heavy. What needs to be realized, particularly by parents (and in these days by critics of the schools), is that this is an essential background without which formal teaching in number or measure is ineffective. Over this same period the child engages in sorting activities, handling discrete objects, and, via sand and water play, finding out what can be done with effectively continuous substances. Gradually the child realizes which properties are independent of one another and which seem related, and how to give names to them.

It is then that the teacher wants to lead the child on to Stage One activities.

Primary activities – Stage One

The first stage in using measure is the development of the concepts of length, size, heaviness, capacity, and duration, without any quantitative associations, together with the comprehension that a large space can be filled or a heavy object can be balanced by a number of smaller objects. These objects then act as a measure of the larger or heavier quantity, but the idea of a unit of measure, and still less that of a standard unit is not yet present. We would not recommend teaching the use of units without an earlier run-up by means of activities which help measurement to emerge as the refinement of a process already familiar. These activities are to be seen by the children as play sessions, although of course the play is highly structured. Most of this work is meant to be done prior to reading and clearly assignment or work cards are not intended. The general aim of all the activities at this stage is to provide structured background experience, from which the later concepts of measurement can emerge. This experience must be offered to children in school because not all of them can get it from home. The children are told verbally what to do, as a group or however the teacher wishes. The child is also learning how to follow instructions given in the more formal modes of speech.

1   Filling a length: working between limits

The child is asked to fill a length with assorted objects, but is otherwise left to do the task without suggestions. Afterwards children should be asked to say what they have done, so that they begin to respond to more formal modes of speech. Children will not always arrange the objects in a straight line, but it is worth noting that they never let their chain of objects go off the paper.

Give children a sheet of paper or card and make, or get them to make, two marks near opposite ends. Ask them to fill the space between the marks with small objects taken from the class sorting boxes or from a reserved collection of convenient articles such as assorted counters, bottle caps, old postage stamps cut from postcards, foreign coins, acorns, conkers, and the like.

Typical questions
>
> Where did you start?
>
> Where did you finish?
>
> Could you make a longer train between the marks?
>
> Could you make a shorter train between the marks?
>
> What would you do if I asked you to fit in more things from the box?
>
> How would you do it with less?
>
> Can you make a train from corner to corner of the paper?

Always note what the children actually *do*. Do they allow overlaps or avoid them? Do they all start from one mark and progress steadily to the next? These activities give us an insight into the way the children are thinking.

2   *A matching game: identifying sequences*

Since the first train or line will be a sequence of miscellaneous objects, the child will need to pick out objects along the line one at a time, identify them, and match up their position on the paper.

With the children working in pairs, get the first to complete a chain of objects as in Activity 1, and ask the second to match it up *exactly*. Do not expect the result to be very exact! It is not necessary to check every result, but if there is an obvious mismatch ask questions using words such as 'same' and 'different'.

> What has Rita got next to her blue bottle top?
>
> What have you got?
>
> Is yours the same, then?
>
> How is it different?

If a match is impossible because all objects of a kind are used up, the child should be able to explain the difficulty.

3   *Filling a length: a move towards abstraction*

Here the objects are representations of other objects less convenient to use. Note that the process corresponds to true measurement, in that the set of objects could be used to transfer the measure to another place – if, but only if, the shape of the line is preserved.

Draw round hands, feet, or lower arms on sheets of sugar paper and cut them out. Children need not do this very accurately: all we need is a collection of plane shapes related dimensionally to the parts of the body covered. Use a number of these to fill the space between two marks on the floor, as for Activity 1.

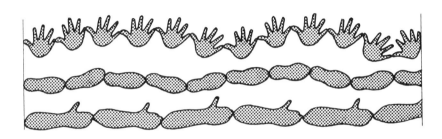

Ask such questions as

Could you do this with real hands and feet?

Can you see which of you has the biggest feet?

If the children are learning to recognize their names when written, the teacher can label a few of the cut-outs when supervising the activity.

4   *Fitting in people: towards estimation*

Although children will probably enjoy doing this, the object of the exercise is serious: it is to compare intervals and objects by eye.

Arrange a suitable interval on the ground between two chair backs or jumping stands, and ask children to guess how many of them could stand between the two marks representing limits of a measure. Tell them to remember the numbers they have guessed. It would be worth while to ask them to say them aloud together: children will at least hear their own voices! Choosing one of the estimates, get that number of children to come forward and try to fit themselves in. By choosing larger or smaller estimates useful questions arise:

Could you do it by standing sideways?

Could you do it by putting out your elbows?

How many more could we squeeze in?

The work can be extended by marking out a square or putting a large hoop on the ground, repeating the call for estimate followed by trial.

5   *Comparison of lengths: linear order (i)*

The difference in length between pieces of string is brought out by threading perforated objects on them and comparing the results. Here the linear order is fixed once the chains are assembled, and the assembly provides a motive for comparing the lengths.

Have ready lengths of flexible string with a button knotted into one end. Ask the children to thread them with buttons, beads, cotton reels, lengths of old ballpoint pen barrels sawn up, and so on. Many families accumulate boxes of old buttons and will contribute handfuls if asked. The children should leave enough string to handle the lengths easily. Lengths should at first be compared in pairs to see which is the longer and which the shorter. To compare more than two lengths requires a concept at a higher level. Do not arrange the lengths in order as a teacher-directed activity, but encourage a group who may be doing this spontaneously. Children may be asked if they can count all or some of the objects threaded. Discuss informally what happens if the strings are pulled out straight, asking whether it alters the number of objects.

6 Comparison of lengths: linear order (ii)

This differs from Activity 5 in not using a line fixed physically by a thread, and sets the problem of comparing lengths if the marbles are not laid out in straight lines from the same starting mark.

Have two jars of assorted glass marbles and ask which makes the longer line when the contents are laid out side by side.

Fill the jars by trial so that they give a satisfactory outcome. If they are very different the result is obvious, but if they are too close it may be difficult to make an easy comparison even by setting out the marbles. By careful choice the experiment can be arranged so that the result becomes quite clear, but only if a starting line is arranged.

Note that we are comparing the intervals filled by the marbles, and not their number. Children who can do so should count them as they put them down; they are then doing a double task. That a longer row may have less marbles need not be made explicit at this stage. This activity is only possible if the children have a level table to work on. A cloth helps to keep the marbles in place.

7 The contents of containers: the bounded space

Although it is hard to image any child arriving at school without knowing the use of words like 'full' or 'half-full', we aim here to extend the experience, linking it with estimation and extending perception by using various kinds of fillings, solid and liquid.

Have a collection of selected containers such as plastic beakers, paper bags, and cartons with boxes of assorted small objects. If those from earlier activities are used and the children are sitting around tables in groups, it is (just) possible to keep the different kinds from being mixed. Get them to put out a pile of objects on the table top that they think will fill the container given them, and then check by trial.

Link the activity with the usual class 'water play', by asking children to put as much water into a jug as they think will just fill a plastic beaker, checking by trial.

Extend by discussing which containers can be filled with water and which can only be filled with solid objects. Have them fill containers from a box of acorns, a sand tray, a jar of rice or dried peas. Suitable leading questions (often needed in teaching) would be

Can you pour the rice like water?

This plastic beaker has a crack in the bottom. Could you find out how much it will hold?

Could you fill it with rice?

Note that the 'how much' in the question is not expected to be quantified by the answer. Filling with small grains is, incidentally, an established technique to measure a container that will not hold water or must be kept dry. Palaeontologists use poppy seeds to measure the brain capacity of fossil skulls.

8 *Towards Archimedes: the displacement of water*

Every child who has ever had a bath must know that the water level rises when you get in, although the experience may never have been put into words and discussed. What is less obvious is the upthrust of the water. This is a background activity only, and we do not recommend attempts at explanation. It is a pity that many interesting phenomena are only investigated at school when the children are mature enough for precise interpretation of the results.

Get children to sort objects into those that float and those that sink. If a child notices that some things float but sink when filled, then of course he is encouraged to talk about this. Have plastic tubs floating in a bowl of water, and fill them with stones till they sink. Extend the activity by blowing up a carnival balloon and letting a child try to push it down to the bottom of a bowlful of water stood in a sink. If someone asks why this is so hard to do an adequate response is:

Let me try. Yes, when you push it down it feels as if the water is pushing it up.

Note that this is still an activity in measure. The stones put in the tub measure its buoyancy and so does the force we have to exert on the balloon.

9   *The perception of weight: density and volume*

This is given as an activity separate from 7, but could be combined with it since the same apparatus is used. Every child must have lifted light and heavy objects, but has probably never discussed the experience systematically.

Using boxes and bags of the same size but filled with different materials, note the difference in weight. Questions could be

Which box is easiest to lift?

Which of these two is the heavier?

Which is the lighter?

Why is this box heavier than that?

Note that here we are talking about weight, not mass, judging the muscular force necessary to support the object against gravity. We are now asking the children to be more formally discriminating, and the work is beginning to pass into the second stage.

10   *Filling flat spaces: the bounded plane*

Activity 4 has already adapted the work with linear intervals to areas, and we now extend this, although the children are not yet ready for work in tessellation. Give out rectangles or squares cut from stiff paper and get them covered with gummed paper squares and circles. Although the squares will tessellate leaving no gaps, young children could have difficulty in doing this: they may well leave gaps and overlappings. If you point to part of the surface covered with circles and suggest that there is a gap, they will happily stick another circle over the space, overlapping the circles all round. This is background experience, not formal activity.

Clearly the work links up with all-over patterns and collage in art. Ask questions such as

Does your pattern cover all the paper?

Can you make a pattern that covers only half the paper?

11 Marking the passage of time: recurrence

In these next four activities time is seen to have several aspects which are psychologically distinct. Only the adult is able to unify these into a single concept. This, however, breaks into its original parts as often as we use it. For the child, the distinct aspects need to be handled separately, and at first are simply part of the life of the class.

The first aspect is the recurrence of events. Infrequent events such as birthdays or Christmas are probably not thought of by a very young child as 'timelike' at all; but only in terms of memory and anticipation. Nor for that matter are meals, whose incidence is associated with activity in the kitchen and feelings of hunger. Bedtime, however, is quite different. It is usually unwanted, it recurs daily, and is, in many families, linked with a clock. 'When the two hands are together at the bottom, it is time to go to bed.' Such statements, and the related 'No, it's not time to get up yet', link the word 'time' with entirely subjective unanalysed emotions. The first aim of school can be to link and extend.

For most children this is the first time their lives are being more closely regulated by the clock: the teacher announces 'time for play', 'time for. . . .' There are times to go home and times to be at school: the child's life passes in a cycle of daily recurrence. Birthdays among classmates, which most infant teachers announce as they occur, and the other cycles of Friday afternoons, Monday mornings, the ends of term, the changing of the seasons, should all begin to link in with the daily cycles. Gradually the child becomes aware, through this recurrence of events, of the cyclic mode in which we perceive time.

12 Cutting up time: the interval

Until the child can read a clock and has a clear concept of hours and fractional parts, lapsed intervals of time (between one meal and the next, or between the start and finish of an activity) – intervals, that is, entirely occupied by doing other things not devised merely to mark the passage of time – can only be vaguely appreciated. They are long or short as the activity in hand is dull or exciting, and for the child that is the end of the matter.

In the classroom, again, one hopes that the normal use of language is directed, like so much of what the infant teacher actually says to the children, to the ultimate end. The difference between 'Hurry up!' and 'Hurry up! it's nearly story time' is conceptually very marked. Although we have numbered this deliberate use of language as an 'activity', it is one for the teacher and not the child. It is not the child who will say 'We have plenty of time to do that', or 'There is not enough time to do this'.

13   Observing time: a start to measurement

At this point, planned activities once more are needed. When we speak of the time taken by a happening we call attention to it. It is long or short quite independently of our subjective impressions. For the child, we need situations in which something can be seen or heard to be happening as the interval being considered passes.

Have set up for use as required as many timing devices as possible, chosen to mark the passage of time by sound or movement. Call the attention of the children to the device, perhaps getting them to clap or even count as the time elapses.

Here are some examples:

A funnel (or tundish) through which sand trickles into a container.

A long swinging pendulum. (It should be long because the slower motion more easily holds the attention.)

Coloured water from a punctured tin filling a glass or transparent plastic vessel.

A large stop-clock with a sweep second hand.

Note that this activity is to call attention to the lapse of time, and not yet to measure it. The second hand of the clock moves, at first glance, smoothly round the dial, and the division into 60 units is arbitrary. One can say that when the hand has gone right round a minute has gone by, but it is not certain that in this way a child begins to grasp the concept of 'minute' at all clearly; the activity lengthens it in the mind. A pendulum one-metre long will beat seconds closely enough for our purpose, so children can count to this interval, but we regard further discussion as Stage Two.

14   *Time as continuum: the ever-rolling stream*

We also need activities that mark the passage of time when it is not being attended to, chosen to mark intervals of a few hours, a few days, or at most a few weeks. Infant pupils have no grasp of a year as an interval: this is not surprising if it is seen as something like one-fifth of their entire effective lives as free agents.

Set up situations that allow temporal changes to be seen at any convenient intervals, as a continuing process. As before, there is no attempt at measurement or analysis, but the child is absorbing experiences on which the later concepts depend.

Set up a shadow stick in the playground and mark, occasionally, the end of the shadow. The marks can of course be made to correspond to other intervals – clock hands on the hour or ends of classroom sessions – but nothing need be done except to notice the movement. Close a blind or curtain in the classroom, and allow a shaft of sunlight to fall on the wall or floor. Mark the spot and try to see the movement.

Notice the growth of a selected plant or seedling. The familiar bean in a jar is suitable, while a sycamore seedling, held similarly by a blotting paper tube in a laboratory measuring cylinder, sends down a long root fast enough for daily observation. Some children may have grown their own plants and waited for results, but our aim in many of these 'domestic' activities is to compensate the less fortunate.

Mark off the passage of days and weeks. The usual wall calendar should be in every classroom so that the pupils may become familiar with it, but the best way is to have a long vertical strip of paper set up and headed 'today is'. The day and date is written underneath, and is crossed out the next day, whose entry goes below. Many classrooms use a day and date indicator which is changed every morning. This gives the necessary practice in reading day and date, but does not so vividly mark the passing of the school term.

15   Weight: muscles in action

The concept of 'heaviness' is independent of what we shall later consider as 'mass'. A thing is 'heavy' if it is difficult to pick up or offers an unexpected resistance to one's muscular effort. It is this effort that is the starting point of all later work in more difficult mechanical concepts. We all learn to correlate our visual impressions with the kinaesthetic sense that tells us the strain on our muscles, so that in the end we *know* that certain things are heavy and others light, without trial. (Psychologists know that this correlation with visual appearance is easily upset by using trick apparatus.) Things are heavy for two reasons, because they are made of material of a high density or because they are large, and we need activities that point to this difference.

Have a set of closed cartons or boxes of different sizes, empty, filled, or partly filled to provide a range of weights. Ask children, without letting them handle them, to say which is the heaviest and which the lightest. Then let them pick up the cartons, one in each hand, asking them if they were right in their guesses. Question:

What do you have to do to find out if something is heavy?

The child who is used to kitchen scales at home may say that things can be weighed on the scales. The obvious response is: 'Suppose you haven't got the scales.'

If available, a few large cubes or blocks of expanded polystyrene as used for window displays in shops are worth the trouble of getting. Their apparent weight when picked up can be compared with that of strong bags containing several kilogram of sand.

Children should be asked to compare the weights, by lifting, of pairs of objects, using the words heavy and light, heavier and lighter. At this stage we suggest that more than two objects requiring to be ordered should not be used. If a group of children show signs of ordering a set of objects by weight, let them go ahead. We ourselves reserve the process for Stage Two.

Evaluation checklist (i)

In this first stage very little has been done that a child from a lively home would not, in some form or other, have done already. At this age the home background is an important factor. For some children it may become less influential as they grow up, and what was a good home for a toddler may indeed become oppressive for a teenager.

The teacher's main concern is the child who has missed out on a background of childish activity. There may be several ways of investigating this, but the easiest and most direct is to look at the child's use of language, especially when confronted by a simple task or situation to be observed and described. We repeat that the correct use of a word in a context appropriate to the circumstances of using it is almost a guarantee that the concept is complete at the stage being considered. The child who cannot lift a box 'because it is too heavy' may never have handled a pair of scales, but is at least formally aware of differences in weight. The toddler who struggles to move a heavy object is not.

We begin by giving a list of words whose use is to be checked. They are not arranged in alphabetical order because doing so suggests that the list has to be worked through in sequence like the items in a test. A teacher may feel that some could be omitted and that others not there should be included; if so alter the list. Our Stage One is to establish the concepts of heaviness, of passage of time, of bounded spaces, and of capacity in an efficient form from which a second stage towards measurement will emerge. Any difficulty in using these words reveals either a deficiency in experience or the interrelated lack of a fundamental vocabulary.

We give this list at the end of Stage One for convenience, but it should be clear that the task of evaluation should be a continuing one and will go on throughout the stage. At the end of the stage the results of the evaluative checks can be recorded as on page 25.

Some of these Stage One words such as 'small' or 'large' can hardly be missing from the normal child's vocabulary, but we give them because they are as much part of the language of mathematics and measure as 'area' or 'volume'.

Not all of the words in the following list will appear in class reading schemes, but they must all be known and used freely if the activities are to be worked at and the conceptual build-up completed to the first stage. We suggest making flash cards for those words not covered by the scheme the children use. We also take it that comparative and superlative forms will be known where appropriate.

The words are not, however, given to be read, but to be said and used in context. At this stage all instructions are verbal. Work cards or activity sheets are not suitable. There may be a few pupils who could read

here	start	begin	next
there	finish	end	over
to	corner	same	across
from	side	different	along
long	wide	heavy	hold
short	deep	light	enough
far	fit	low	shadow
near	edge	high	soon
large	up	pattern	late
small	down	compare	early
few	even	narrow	fast
less	uneven	broad	slow
more	fill	full	today
between	cover	empty	tomorrow
season	after	before	yesterday

them, but the task of interpreting written instructions stands between them and the aim of the activities.

At this level evaluation is largely observation. The teacher tries to use the less common words in discussing the children's work with them, and tries to get them to use them in response. The technique simply depends on thinking before speaking. Consider, for example, Activity 1. One would normally say to a child: 'Make a mark there', pointing to the spot. If instead one says, while pointing: 'Make a mark near the edge of the card, there' then the two words 'near' and 'edge' have been used. If the child is now asked: 'Where do you think I want you to make the next mark?' the answer will probably be simply 'there'. The teacher can then say 'Tell me where "there" is, if you put the first mark *near this edge*.' There is a good chance of the expected reply 'Near this edge', accompanied by pointing.

Such exchanges go on all day in busy reception and infant classrooms: the only difference – but a key one – is that the teacher is deliberately more precise, filling the normal ellipsis of casual speech with words rather than gestures. We hope that the use of language in this way will spread from child to child, so that they acquire what is to be the working vocabulary of mathematics and measure.

Only children known to have difficulty in keeping up with their fellow pupils need be individually questioned. The demand 'Tell me what you are doing' could well dismay the slower child who is not really sure, and more leading forms of questions need to be devised. There is, at this stage, no question of compiling a long list of standard questions or diagnostic instructions, although a few may be useful. If a child is given a small card and asked to put a mark in one corner, then the response certainly shows whether the word 'corner' is understood in context.

The time spent on the first stage of the activities is a matter for the judgement of the teacher. The stages do not represent a norm for year groups, but levels of mental development that children will reach in their own time if they can react actively enough with their classroom environment. Certainly the two early stages are not dependent on reading skills, and can be evaluated by the judgement of the experienced teacher as effectively as by standardized tests.

Primary activities – Stage Two

Here the work takes a definite step forward. We are trying to develop the idea of a unit of measurement, although not necessarily at this stage a standard unit. This is restricted at first to length and capacity. The concept of area is advanced by tessellation activities, of volume only by an extension of the early work, since it is still far too soon to think of units for these quantities. We shall begin to associate the activities with number skills as they are acquired and we shall make sure children are becoming familiar with clocks, at least to read the hours. We shall also introduce very simple graphical representation where it correlates with measurements such as length. There is a corresponding step forward in richness of vocabulary which will allow evaluation of this stage to be done rather more formally.

16   *Introducing units: making up lengths*

Here we make up a length using, rather than objects chosen at random, a selected article that can act as a unit. The work can be extended by counting and the use of display graphs.

Although it is now not so easy to collect large numbers of matchboxes, one can usually get sufficient numbers by sending a request home with children. Crown bottle tops are easier to get; and any pub will supply a bagful for the asking.

Choose a number of convenient lengths, such as distance across a table or along a sand tray, then have them filled with matchboxes placed end to end. The matchbox is ideal in size and shape for this purpose. Ask the children to count the boxes.

The work can be repeated, choosing lengths that are convenient for the units available – milk straws, bottle tops, or coloured arithmetic rods. The use of these last is important, since it will link up with metric measures later. They are used without comment at this stage. Consolidate and extend the work by graphical representation, with actual matchboxes or bottle tops stuck into position. The graph needs to be made on a large sheet of sugar paper, the teacher at first setting out the work and a few children sticking on the tops or boxes. The work may introduce new words that the pupils have not yet read: this can be dealt with then and there.

Finally, the word 'measure' can be introduced as a verb describing what has been done. We measure out the sand tray with matchboxes. If coloured arithmetic rods are in use, the term will already have been introduced in 'measuring' one coloured rod by another. This, of course, is a parallel activity and each reinforces the other.

17   *Frieze patterns: repetitive units*

Here a motif is repeated as a pattern to fill an interval. It is hoped that the children will enjoy doing it as a decorative project.

Get the children to draw round their own hands or the feet of partners standing on sheets of paper and then cut out the outline. The children should work as neatly as they can, and there is no harm in getting a piece of work repeated if it is well below the known ability of the worker, but we are not after exact outlines. The feet or hands, which can be called foot or handprints, may be coloured or crayoned. The cut-outs are then stuck as a frieze along a wall of the room, on a long strip of paper (odd rolls of cheap wallpaper are ideal) put up for the purpose.

Discuss the situation:

How many hands does it take to fill from here to here?

Have we got the same number of feet?

Do the feet stretch further than the hands?

This piece of wall is 27 hands long. Why do the feet stretch further than the hands?

There is also a very important leading question:

You have got a lot of spaces between the feet. Does it make any difference if we put them carefully together?

And the comment:

Let's do it that way so that we can see just how many footprints we can put in without putting them one over the other.

Note that these units are only very approximately constant: this point need not be raised with the class.

18   *Balance: comparison of weight*

This activity introduces an early piece of apparatus, the equal arm balance. These are available for infant classrooms as play ma- terial, with large dished pans. They are usually insensitive and not adapted for accurate measure, but they do what we want better than more sophisticated devices the children may find hard to handle.

Get a child to say which of two objects (chosen for their disparity) is the heavier, before suggesting that they are put one in each pan of the balance.

What happens to the heavy one?

Try changing them over. Does the heavy one always go down?

Have the children compare pairs of objects in this way, until there is a chance for the question:

What happens if you put on two things that are the same?

Introduce the verb 'to weigh'. When we put things on a balance (or scales) we are weighing them. The one that makes the pan go down weighs more than the one that comes up.

Extend the word by discussing this situation as far as possible, using the balance for a wide range of sorting and, eventually, numerical exercises.

For example:

Sorting into 'heavier than', 'lighter than', and 'the same as' a given object.

Experience in balancing materials of different densities, against one another or against the same object – flour, rice, acorns, dry sand, wet sand.

Straightforward number work, balancing or comparing equal or different numbers of uniform or fairly uniform objects – cubical building blocks are ideal.

Balancing objects against others after guessing what numbers will give equality.

Children can also try balancing in other situations, such as with a lath over a chair back. Compare this to balancing oneself on one foot, or on a P.E. form. Setting up a class mobile using fishes or birds made by the children is worth doing: adding one figure tilts the structure until it is adjusted, that is, it 'weighs it down'. This, of course, is an example that may lead into difficulty, since one can either add a figure to the other side or shift the balance point. There is no problem if one uses only the terms 'weighs down' and 'balances' or 'out of balance'. There is no reason at all why children should not get the hang of balancing unequal arms by choosing a suitable pivot point, but the situation would merely be accepted and not extended formally by question or discussion.

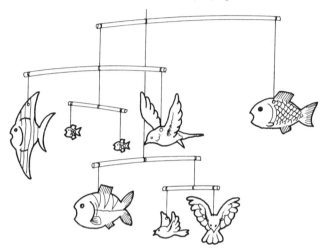

Note that *all* measure is comparison. We compare some aspect of the physical world with something selected as a unit, and decide whether it is quantitatively less, more, or equal.

19   *Filling bounded planes: tessellation*

In tessellation work (see *Children Learning Geometry*) we are usually aiming to show that a tessellation extends indefinitely in all directions. Here we use bounded shapes or outlines, which are to be compared by the number of tessellation units needed to cover them.

Children can produce rectangular shapes by drawing round boxes or cartons held flat on sheets of paper. They are then asked to fill the outlines with gummed squares or other shapes that tessellate. If we want to know exactly how many fit in, we must avoid gaps and overlaps as far as possible; at first children will have difficulty with this. Later, they can be asked to guess the number of units they will need.

Hand or footprints provide irregular outlines for unit tessellation with small squares. At this stage ask the children to do their best to cover the

shape, but do not stress the formal requirement of dealing with the irregularities. It is possible to treat the whole exercise as one in pattern making, using coloured units of gummed paper. The only formal condition is that the children use filling units of the same shape for any one outline.

20 Covering surfaces: copying areas

This is an exercise to call attention to the amount of surface presented by a small box or carton.

The children draw round a box to make its 'footprint' as before, but then cut this out and stick it on the face of the box. They then repeat this with the other faces. In the end they have covered the entire box, and have seen how much paper they have cut out. With children who are making good progress it would be worth asking whether it is necessary to cut out each face separately: is there a quicker way?

21 Curved surfaces: judgement of shape

This activity requires the child to judge which of a number of possible shapes will fit round a cylindrical surface. It is worth watching anyone set this task: if one of the correct sheets is picked out without trial, the child probably has reasonable spatial perception. If both correct sheets, outstanding perception is likely.

Obtain an empty drinks can or similar cylindrical container. Make up six shapes from thick strong paper, numbering or lettering them for identification:

1   A rectangle of the correct dimensions to cover the tin or container, allowing a small overlap for pasting.

2   A rectangle too long but not wide enough.

3   A rectangle too wide but not long enough.

4   A rectangle, correct length (with small overlap) but not wide enough.

5   A rectangle, correct width but not long enough.

6   A parallelogram of the correct width and length, allowing overlap.

Ask a child to pick out which of the numbered sheets could be used to make a label for the can by winding it round and pasting the overlap. It is quite likely that a child will decide on the first sheet and then abandon the others. If so, ask for the rejected sheets to be demonstrated then and there. This is a valuable activity in that it produces, for the child, an unexpected result. A pupil could be asked whether the wrapping of the labels shows that there is as much paper in the rectangle as the parallelogram.

Note that the word parallelogram can be introduced in context as the name given to the shape in question. A full discussion of the use of 'technical' words in mathematics is given in *Children Learning Geometry*.

22   Ordering: bigger or smaller

This is a simple task that can be set at any time, and if well prepared can give practice in using letters of the alphabet.

Prepare a number of 'ordering boxes', small trays or open boxes containing cardboard strips of the same width but of various lengths, or shapes such as squares or circles of different sizes. Label them a, b, c, d . . . in any order. The child is asked to arrange them in order of size, starting either with the smallest or the largest, and then write down the label letters in the order that emerges. The teacher will need a note of this. Numerals can be used in place of letters, allowing children to see that numbers do not always have to be in 1, 2, 3 . . . order. By extending the number of pieces in each tray the task can be made more difficult, since the sorters have to fit in more pieces between those already put down. The work itself can be extended by making some of the strips or circles only slightly bigger than their predecessors in the sequence. The children will then need to compare them carefully by putting them together.

23 *Construction of volume: the building unit*

This is one more activity that many children will be used to from a very early age. Eventually we shall need to build up cuboids (or oblong blocks) from cubic units, but for the moment all that concerns us is that certain shapes can act as building bricks for solid structures.

Any infant classroom will have a good supply of building blocks. The sets of blocks sold as toys for actual model buildings, where the shapes are assorted to allow various features such as arches to be built, are less useful for mathematics than sets of cubes. Sets of coloured arithmetic rods are also useful. Ask the children to build up, using the blocks, the shape of a given box or small carton, as nearly as they can. Since the chance is small of being able to build up a shape exactly, or even to fill an open box completely, using solid units, this activity is not to be regarded as equivalent to the work on length or mass. We need it as background experience to which the children will return later.

A final extension of the work is to ask how many differently shaped solids can be built up using a given small number of cubes. It is much better to have this task done with a good supply of cubes, so that each structure can be preserved for display and discussion. With 128 blocks a number of oblong blocks (or 'cuboids') can be made: $8 \times 8 \times 2$, $8 \times 4 \times 4$, and so on, but this is rather finicky work and should be deferred till a later stage. With 6 blocks a number of interesting shapes can be made and discussed: at this level the question whether two shapes are the same or different is entirely open-ended and is best left without a conclusion.

What is important is that the children should answer without hesitation the questions:

Do these two shapes take up the same amount of material?
Would these two shapes weigh the same?

24 *How shapes grow: a counting exercise*

This activity will recur at a much higher level, but it can be presented here. We should always take the chance of using more advanced topics in an elementary way.

Provide the children with a box of identical squares and ask them how many squares have to put together to make a bigger square. Someone who suggests nine should be asked if it is possible to make a smaller one. Now make larger and larger squares, counting them and recording the numbers.

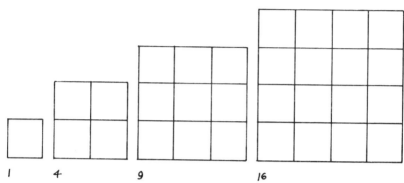

1   4   9   16

If the squares are later stuck into position, a wall display can be set up, under the heading 'how squares grow'. The sequence of numbers is written underneath, but they are not (at this stage) discussed as 'squares', although they can well be read out by a few children aloud in sequence. If this activity has gone well, it can be tried with a box of cubes, although it is even more important to let the pattern of numbers go by without comment.

25   *Sequence in time: the daily programme*

Many infant teachers include daily activities in the early stages of word recognition and reading. If these are dealt with in time sequence using the words 'first' and 'then' the exercise has a double use.

Every teacher will have an individual approach to this sort of work. What we have in mind is a list of sentences taking the child, in a few or many steps at the teacher's discretion, through the working day.

We come to school

First we take off our shoes

Then we sit on our chairs

Then we go in the hall

and so on.

Certain regular activities such as the taking of milk or having story time can later be associated with clock hand positions.

26 *Capacity: towards a measure*

This activity repeats what has already been done for length and weight, but because of its simplicity the litre measure can be included among the unit containers used for capacity activities.

Get children to count and record the filling of larger containers by repeated use of smaller ones. If they can now read simple incomplete sentences like

The jar A holds . . . . plastic cups of water

they can be given these on work sheets and asked to fill the gaps. If they can write them directly (or from a master copy if the teacher works in this way) so much the better.

Note particularly that 'jar A' is *not* a technicality like 'the triangle ABC' in geometry. The jar is to be clearly labelled A so that it can be identified. It is called the 'jar marked A' and then, by ellipsis, the 'jar A' without comment. If a word such as 'plastic' is not in the reading scheme, it should be introduced as any other word familiar to the child through use although rarely seen in reading matter. Words which describe articles in common use in the classroom should always take precedence over those that need pictures to give them substance.

Include the litre jug among the vessels supplied, asking how many cups fill the jug and how many jugs fill the bucket. It is also useful to have containers which are cylindrical in shape, so that one can ask for them to be filled half-full, by eye. The conical container is too deceptive for this purpose. Always take the opportunity to use the word 'half' in context: it is, as we point out in *The Third R*, a bridge between ordinary language and mathematics.

27   *Tension and extension: how things stretch*

This is an activity often used quantitatively among older children, although in fact it tends to give erratic results. The basic experience only is needed at this stage: the greater the weight, the greater the stretch.

Provide lengths of elastic tape and let the children stretch them and feel the resistance offered before they begin the task described. Fasten a pan or an improvised basket on the end of a length of elastic supported from above. It helps if a metal hook is fastened securely at each end so that the children can deal with it from there; otherwise making it fast is beyond them.

A length of stout quartering or dowelling carrying a hook can be fixed permanently to a cupboard top, although preferably a wall bracket could be part of the equipment of a classroom. This can be used for

pendulums, plumb lines, stretch experiments, and the like, as well as being a convenient hanger for charts or pictures under discussion. The pan should be loaded with cubes, or dry sand run in from plastic cups which thus act as a measure. The children establish that as the weight increases the stretch increases, and as the weight is removed the elastic recovers its original length.

Teachers should ensure by trial that the hooks are securely fixed and that the lengths are chosen so that the pan reaches the floor before the elastic is overstretched. It could whip back suddenly if anything broke.

There should be no attempt to discuss the experiment in quantitative terms, although a group might care to mark positions on a strip of paper behind the elastic.

28   Stepping it out: measure with one unit

This is a key activity that children do not do very precisely. So far we have filled an interval with units and have counted them. The work is now considerably extended by using one unit (or two, see below) to pace out a distance. A technique is now called for.

The work is probably best done as a sequence of activities. Make two chalk marks on a table and ask a child to fill the space with matchboxes (i.e., 'measure out' the length). Count them and make a note of the number.

Remove the boxes and give a child one box only, asking if the length can be 'measured out' with this.

Many children would simply pick up and put down the box by eye, counting as they go. A knowing child may use a finger to mark the box end as it is taken up. The number then counted will certainly not be the original number. Discuss this and get suggestions, leading to the second solution of the problem. A light chalk mark will be more satisfactory than a finger on the table.

The question 'Is there another way?' is unlikely to get a very useful response, so hold up two matchboxes and ask if anyone could do it more easily using them both. After the group has arrived at a solution by stepping the two boxes, each marking a unit interval end on with its predecessor, try to get the children to see that this is what one does when stepping or pacing out a distance, putting heel to toe with or without a stride in between.

In Activity 3 we used cut-outs of the hands and feet to fill spaces. It is often suggested that children should measure using hand spans. In spite of the apparently egocentric motivation, we do not recommend this, because when 'stepping out' the intervals with the outstretched hands it is quite difficult for children to get their hands crossed over without twisting and swerving away from the line of measure. The illustration to Activity 3 shows what usually happens: as a pattern this is all right, but the zig-zag line is better not introduced into measure.

29   *How hot is it?: introducing temperature*

Although work with thermometers is better delayed until children can safely *use* one, reading temperatures can be done much earlier and we suggest that the first part of the programme should be started as soon as the pupils can record, as a Stage Two activity.

Have a wall thermometer in the classroom and take the temperature, preferably at the same time each day. Have a notice reading 'today's temperature is . . .' and prepare a set of cards giving the expected range so that one can be selected and inserted each day. Show a few children how to read the thermometer – a large one made for class use with no great expectation of accuracy is what is needed – and get them to show one another. It is worth persisting with this even if pupils are slow to succeed. They can be told that the reading is zero when it is beginning to freeze and would be 100°C if it were as hot as boiling water, but nothing else is needed until Stage Five is reached. Many children will hear temperatures given during broadcast weather reports, so that this activity fixes what they hear in context. It is also likely that they will begin to associate readings with the actual conditions, and will soon be able to estimate normal temperatures.

The work can be extended very usefully if another wall thermometer can be placed outside the school for comparison. This must be hung on a north wall where it is out of direct sunshine: otherwise it absorbs heat from the sun and does not register the air temperature.

Evaluation checklist (ii)

Although neither Stage One nor Stage Two has yet reached the level of applied mathematics, it should be clear that they represent a progression. They pass from the recognition of objects as larger or smaller and the use of the smaller to fill out the larger, to the fitting in of equal units and the concept of ordering. There has also been a link-up between the activities and the child's developing skills with number and words.

As a result, evaluation can be more direct. One can devise situations in which the use of words in a practical context can be checked. The previous list of words will be kept in constant use and the following are among those that teachers will need to add. The gap between words taught for literacy and those needed by the children to cope with other subjects must be kept small if mathematics is to be learnt. Any 'special' words on the list are only special in that we can usually do a lot of reading without them – but very little mathematics.

pace	equal	stretch	hold
stride	half	whole	unit
enough	nearly	level	circle
height	check	overlap	elastic
width	match	overflow	square
measure	cube	triangle	balance

As before there is no suggestion of an organized test for these words in action, although no doubt one could be devised. Whenever there is an opportunity, with the necessary material to hand for the child to manipulate, a question can be put or a simple task verbally set that shows if the child has an operational grasp of any word selected.

We can give some possible tasks and questions appropriate to a few of the words given. In spite of the trivial nature of some of these, it is the total experience that counts.

Stretch this piece of elastic till it is twice as long as this piece of stick.

How many of your paces are there between here and the door?

Find out who has the longer stride, you or Anne.

Take six squares out of the box of squares and put them down in a line so that they overlap a little.

Put a circle on the table and find a circle that will hide it when you put it on top.

C.L.T.M.—E

Measure the height of this table with a pile of cubes.

Show me whether this jug will hold four cupsful of rice without overflowing.

Note that we do not try to distinguish between circle, circular shape, and disc. It would be a good group exercise for in-service teachers or students to make a working list of evaluation tasks and questions. Except for a very few tasks, undertaken by children whose reading age is well ahead of average, assignment cards are not suitable, and should be deferred until the next stage.

Primary activities – Stage Three

Stages One and Two differ only in that the second is more systematic and moves towards the quantitative. In Stage Three we actually begin the process of measurement with standard units. We have given a lot of space to the first stages. Although they are preliminary activities, they are quite fundamental to what we are trying to do. Measurement is a skill, calling for manual dexterity and mental judgement, and this skill can only be developed by practising it. It is only marginally connected with 'doing sums'. The teacher, not the calendar, will decide when the child is through the stages. Nothing should be rushed. There is no educational reason why children should begin a systematic approach to Stage Three before their second year in the junior school, although no doubt some children will wish to do so. Seven-year-olds are a long way from measurement in order to cope with life and earn a living, but they are beginning to need a vocabulary to deal with size and distance and time, and a set of skills with measure that gives substance to these words.

If the work of the earlier activities has been done well, the third stage should go easily and be fairly short: the basic skills are there and are now being applied to the formal units of measure. The new operative concept that will be introduced is the *fraction*, the mathematical device that makes measure by standard units possible.

We should expect many children now to be able to read the words on the evaluative checklists, and simple assignment cards are possible. The work now being done in measure will keep pace with number skills. Measure can be thought of as number in action. Apart from straightforward counting of people or objects and the special use of number in handling the currency, there are few applications of arithmetic that do not involve measure.

30   *Thinking in litres: measuring liquids*

We want children to think in terms of the litre. Not many adults are able to think in terms of the old imperial units, except for those that happen to impinge on their lives. With SI units, however, the one unit is extended to every possible application, so that the work done in familiarizing will remain useful for life. We begin with the litre.

Have jugs or other measures in litre, half litre, and quarter litre capacities, and a litre measure graduated in halves and quarters. If purchased equipment is marked in millilitres, showing 500 m*l* and 250 m*l* instead of the fractions, cover with small labels stuck or (with polythene containers) sellotaped into position. Allow the use of these to form a standard whereby the original haphazard collection of jars and jugs is reduced to order. Find out how many cups or beakers fill the measures up to the mark, and how many of the measures are required to fill bowls or buckets. Accompany the work, as always, by guessing or estimating.

Those who want school leavers to be able to handle measure must consider how often we take a rough estimate of a measure rather than a careful determination. This process not only promotes reasoning, but is even more important in adult life, where the choice of a measuring process so often depends on knowing what sort of result is expected.

As an extended activity it would be interesting if water could be added to containers of dry material such as sand, to see how much can be poured in: this, however, ruins the dry materials as such. It could be organized as a class lesson using a small bucket of dry earth, sand, or gravel, allowing each child to pour in a plastic cup of water until a puddle is formed on the surface. The result can be recorded:

We poured 21 cups of water into the pail of dry sand.

If, however, the cups are filled one at a time from a litre measure replenished as necessary, the activity then becomes an investigation in standard measure

The pail holds 5 litres of sand.

We poured in over 2 litres of water.

Or, perhaps,

We poured in nearly $2\frac{1}{2}$ litres of water.

It is in this way that the transition to measure by standard units is made, with the use of fractions as an extension of the role of number.

31   *Comparison of heights: linear order (iii)*

Although one would normally compare the heights of a few children by standing them together, this activity, by transferring their heights to strips of paper, takes the work towards graphical representation.

The work is most easily done with rolls of wide paper tape, which in turn are readily and cheaply obtained by cutting with a hacksaw sections of a few centimetres from a roll of lining paper as used by decorators. The children can be given strips of a suitable length and asked to mark out and cut them to match their own heights, and we suggest that the task is left quite open with no more in the way of instructions. No doubt the results will be erratic and some of the methods unreliable, but the class should now have a wide range of solutions to the given problem to compare and discuss, possibly concluding that there is indeed a 'best way' of doing the job – perhaps by pinning up a strip against a wallboard panel and standing against it while another person takes off the height and marks it.

The strips can be labelled and compared, perhaps by pinning up or pasting on sheets of wallpaper. There is, as yet, no question of measurement: what we have done is to match up personal heights against lengths of strip. The strips can be handled, displayed, compared, and ordered. The work is extended in Activity 47.

32   *Larger and smaller: metres and centimetres*

We need to measure small lengths with a small unit and greater lengths with a longer unit. The standard system has only one unit, the metre, but uses submultiples for smaller dimensions. The children will use, at first, only one submultiple – the centimetre.

The relation between the two can be formally discussed later. Since we think the metre and the centimetre should be introduced concurrently, all the early work is included in this one activity, which is clearly one whose work should extend over a considerable period.

Provide children with ungraduated rods marked 'one metre' and get them to measure out distances across a room, along a corridor, or in the playground. It helps them to keep a line if there are floorboards or lines of tiles across the room. They can be told to measure down the middle of the corridor or to keep near the wall of the playground. It is better for the children to work in pairs, using both the two-rod technique (familiar through the matchbox stepping of Activity 28) and marking the end of the rod on the ground before moving on. From the very beginning they can estimate halves, and in this way come to an approximation when the measure is not a whole number of units.

If the children have already met quarters by folding strips or any other method, the work can be extended by measuring in quarters. The concept of 'to the nearest half' or 'nearest quarter' is difficult: leave the matter open. For example.

You need about half another metre stick to finish off with, so let's call it $12\frac{1}{2}$ metres.

The children will probably build up large cumulative errors because of gaps between steps. We shall not achieve great accuracy.

What is most important is to keep a record somewhere that the work has been done. That a child should rehearse skills by measuring a playground is excellent; but that this should be done year by year as the child moves up the school, each time as if it is a project newly devised and thought of, is deplorable. One of the many advantages of having a teacher appointed as mathematics coordinator is to avoid this repetition.

Concurrently with this work, the pupils can measure smaller objects in centimetres. They should have rules about 30 cm long divided into centimetres only (no millimetres!). Ideally, for younger children, the centimetres would not be numbered, but coloured alternately so that they could be *counted*. (The children can, quite soon, be given measures with numbered centimetre graduations.)

Since the children will earlier have measured intervals by filling them with centimetre cubes, the transition should be clear to them if it is discussed with a suitable practical task. For example, give each of a group of children a card cut to an exact number of centimetres. Get them to fill up across the card along one edge with centimetre cubes, either plastic centicubes or wooden arithmetic blocks. Repeat, using rules with alternately coloured centimetres. It is not necessary to begin the measure from a marked zero. Discuss the results and suggest that from now on we shall measure all fairly small distances and lengths in centimetres, as we measure longer ones in metres.

33 *Larger still: the kilometre*

At this stage the kilometre is not a unit that can conveniently be *measured*. It can, however, be introduced so that children become familiar with it.

Most children will absorb, from general reading and conversation, the idea that distances can be expressed in miles. The kilometre can be introduced as the international alternative. By using the large-scale plan of the locality that should be available in all schools, one can easily mark off recognizable places one kilometre from the school. Many children will no doubt pass these on their way to or from school. The class could be taken for a one- or two-kilometre walk.

The unit needs to be discussed: it is a measure got by stepping out a metre one thousand times, and is something like the distance covered in a quarter of an hour's gentle stroll. The work can be informally extended by 'stretching out' the kilometre to a straight line which can be marked clearly at the bottom of the map. For this purpose one can ignore the actual map scale if given: the line should be drawn in boldly or marked on a card pinned on to the bottom.

This line represents one kilometre

or

When you have walked this distance on the map you have walked one kilometre.

No further extension is necessary at this stage.

34 *Making a start: numbered scales*

The use of zero on a numbered scale is a very difficult concept. By watching younger children using the older type of school ruler which extends at each end beyond the graduations, one can readily see that few of them bother to get the zero point properly aligned on what is being measured. We recommend guided instruction using apparatus devised for the purpose, by introducing a numbered scale without a zero.

The children need 15 cm graduated rules with each centimetre numbered, which they can make as a craft exercise from thin card and a strip of graph paper. The centimetre intervals themselves are numbered 1–15. We do not yet use a graduated rule with numbered divisions one centimetre apart. The diagram shows the sort of rule that we have in mind.

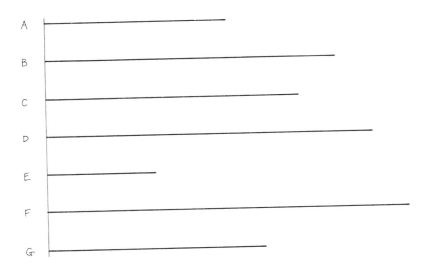

We suggest that the activity should be completely structured. Give each pupil a duplicated sheet of measured lines, each of which is a whole number of centimetres. The lines can be identified, as in the diagram, by letters.

The children will also have a prepared table in which to enter their results. Later, preparing such tables can be a useful part of their work, but we want them now to be able to concentrate on the act of measurement.

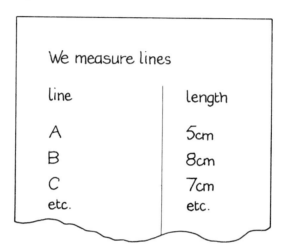

We measure lines

line	length
A	5cm
B	8cm
C	7cm
etc.	etc.

Teachers can approach the task as they wish, either as a completely formal class exercise, as a free exercise after one or two have been done together by the class, or in any other way. The final result is that each pupil should have a table of lengths which can be checked against the known values. Action will need to be taken if some of the results are erratic, possibly by asking the child concerned to do the task again (or on another set of lines) with the teacher.

When the task is complete the children will have had a formal introduction to the use of measure, as it could be needed outside the classroom. We must realize that the accurate reading of numbered scales is not an easy task, although it becomes one after continued practice. It is something that children should be asked to do continually from now until they leave the primary school. During all exercises the children can be asked to guess the measures before they make them: this process should become almost automatic.

A 15 cm rule is chosen because it is easier to handle with the set of lines and later with conveniently sized objects such as soap cartons. It also keeps the numbers concerned down to the completely familiar 1–20.

The rules are conveniently made with strips cut from one-centimetre graph paper. These are stuck along the edge of strips of card, which can be guillotined in advance by the teacher, and cut off neatly to be exactly 15 cm. The squares are then numbered clearly from 1–15 and the whole rule is covered with a doubled strip of sellotape. Allow the children to keep their rules and take them home.

Extend the work by measuring suitable objects, with a very informal discussion, to cope with the fractional parts. As with the previous activity, an informal use of halves can be made or the phrase 'Let's call it . . .'. We do not expect Stage Three work to be very prolonged, and the pupils will soon meet more precise standards. Children can produce neat tables of measurements for temporary wall display, if only as an incidental exercise that keeps words such as 'length', 'width', and so on in use.

35   *Using the balance: the 10 g mass*

The fact that the kilogram is rather heavy and the gram rather light for classroom use can be turned to account as an informal introduction to a multiple unit. This is very convenient for the range of small objects usually weighed by children. Note, however, that although the balance illustrates a very important principle, it is no longer in common use in its simple form.

Have a large supply of plastic 10 g masses. The best ones are discs clearly marked 10 g which are moulded so that they can be stacked to make larger multiples. One would usually speak of these masses as 'weights'. May we suggest both words are used, keeping them in front of the children till the two concepts can be sorted out later? Using the classroom balances, weigh up suitable objects using the 10 g masses. It would be worth while to put the kilogram and the gram (suitably mounted) on display in the classroom, labelled:

This is a kilogram. It is rather heavy.

This is a gram. It is very light. One thousand grams make up one kilogram.

Obviously most children will see the connection between the gram and the ten-gram weights, but this need not be formally stressed. The beam balance in its classroom form is not very sensitive, but using the 10 g units should allow the children to weigh up bean bags and the like reasonably well. A box of articles reserved for weighing can be assembled. Weighing sheets can be duplicated to give practice in writing both the names of the objects and the mass in units. This is, of course, a numerical exercise in counting and recording in tens. One can now obtain plastic centimetre cubes of mass 1 g. These are useful in that children can balance up any convenient object against a counted number of grams, which need not be multiples of 10. Although the scales or balances are unlikely to be accurate to the nearest gram, the exercise allows the children to work nominally to three figures, and can therefore be of mathematical value.

36   Introducing household scales: weighing in practice

Beam balances belong to the archaeology of the retail trade, and the commonest form of weighing machine now found in shops has a digital display. Nevertheless, the older compression or swinging arm balance with a numbered scale is still found in use and is the usual form of apparatus found in kitchens or the house. Because one does not have to do anything with this balance except read the pointer position, it is an admirable device to introduce the graduated scale.

A convenient weighing machine has a scale divided into 5 g or 10 g divisions. Some but not all of these divisions would be numbered on commercially supplied apparatus. Discuss the scale with the class, allowing pupils to alter the zero adjustment that makes the pointer begin properly at the beginning.

Make up masses of 50 g or 100 g using the 10 g units, and show how the pointer indicates their weight. Some children will already be quite familiar with scales of this type, but others will not and must have the chance of experimenting with them. Discuss reading the scale to the nearest numbered division, counting on in fives to get the final weight.

As before, weighing sheets can be used and completed by the pupils. Here, since the actual technique is so rapid and simple, the work can be extended by recording the guessed or estimated weights before each object is checked. The final sheets would look like the table given.

The error can be recorded. This requires the children to find the difference between the two columns. They can be introduced to the plus and minus notation – at this stage merely a device to record high or

Weight of objects

object	guessed wt.	actual wt.	error
Pencil case	150g	124 g	+26 g
Shoe	210g	235g	−25 g

low estimates. Teachers will realize that such prepared tables are only temporary frameworks to help organize the children's activities. It is not intended to force every class activity into a formal mould, and sooner or later children should devise their own tables.

37 Measuring area: the square as unit

Our own modern measure of area uses units derived from length, such as the metre squared (or square metre). This is a difficult concept. The Stage Three work with area will be confined to using standard squares as units, but will also try to establish that area is independent of shape. The work relates to geometrical activities and can, if the teacher wishes, be done at the same time.

The standard units we shall use are 2 cm squares. These are chosen not only because they are a convenient size but because 2 cm squared paper is readily obtainable. A continuing supply of squares can be cut by the children from smaller sheets, or from strips of 2 cm width guillotined off by the teacher. If the children's cutting is not very skilful, strips would be better. As before, we are not aiming at accuracy but at establishing a principle.

It is debatable whether work on area should concentrate on irregular outlines or on rectangles and shapes that can be built up from them. The standard unit suggests the rectangle, and indeed the use we usually make of area in buying carpets and the like suggests that this is what is really wanted. The concept of area, however, is independent of shape. Probably both beginnings can be used with advantage; we shall use both and the teacher can choose.

We recommend drawing round books, boxes, and other regular objects as well as 'footprints'. This transfers the dimensions under consideration to a sheet of paper and allows them to be handled as required. The children should actually stick the squares in position over the outlines, counting them and recording the total.

It is possible to structure the earliest exercise completely by issuing rectangular templates in strong card cut to exact multiples of the 2 cm unit. If this is done, one must extend the work later, either by issuing templates not in exact multiples or by considering the gaps and overlaps left by the class in their activities. Possible questions are:

If we were to take three or four more squares and cut them up, could we fill the gaps, more or less, with the pieces?

If we were to cut off some of these overlaps and stick them in the gaps, would it tell us better how many squares cover the area?

38   *Shape and size: working with constant area*

The independence of area and shape can also be approached from the other direction by using a fixed number of units to build up various shapes. Many activities can be devised apart from the examples given.

Issue or get the children to cut out sets of six 2 cm squares. They are asked to make as many shapes as they can using these six, firstly without any cutting and secondly by cutting some of the squares in halves, diagonally or at right-angles to one side. It might be advisable to demonstrate the two cuts. If the squares are brightly coloured, the shapes produced can be stuck onto sugar paper for temporary class display.

Get the children to cut an agreed number of unit squares, say 25, and stick them on sugar paper to make a chosen picture. Label the picture with the area. For example:

'This man is 25 square units.'

The class production usually makes a lively temporary display. The tangram is an excellent way to stress the conservation of a fixed area as shape changes. The work discussed in *Children Learning Geometry* can be extended quantitatively by beginning with tangrams cut from squared paper in 2 cm units. If the tangram square is 5 units by 5 units the children can see, by counting or by any other method that occurs to them, that the large square is made up of 25 units. The duplicated tangram design issued to the class for cutting up could show the 2 cm grid as well as the lines for cutting, as in the diagram.

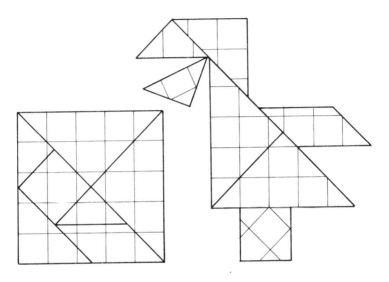

39   *Reading the clock: time as a cycle*

There are, psychologically and practically, three distinct aspects of time. There is epoch time, the continuous passage of days and years from an arbitrary point in history. There is the time interval that elapses between two events, and there is the daily or annual cycle of events that gives us our calendar or clock time.

So far we have tried to get children to appreciate time intervals, and we have got them to note recurrent events in their day in terms of clock time. We now want to discuss telling the time as an activity. Epoch time itself can scarcely be taught: it is a concept that develops with maturity. A later activity refers to it.

The inclusion here of a numbered section on reading clocks is obviously only for convenience of reference: as an activity it is spread out over the child's life at home and at school. Some children, for one reason or another, may neither see clocks at home nor have their activities regulated by them. The school must provide a complete programme, which can be skipped or run through by any one group of children as the teacher decides. The progression we recommend for the ordinary clock is:

Concentrate at first on the hours 1 to 12.

Teach half-past each hour.

Teach quarter-past each hour.

Teach quarter-to each hour.

Count the minutes in fives.

Count in ones.

The work is best done with a clock face from which one or the other hand can be removed to allow discussion of hours and minutes separately. The hands should also be of different colours. Today children will, perhaps more often than not, hear 'ten fifteen', 'ten thirty', and 'ten forty-five', in place of quarters and halves. These equivalents, once heard mainly on railways, are becoming more and more common with the spread of the digital clock and should be taught as soon as the children can count minutes. Such forms as 'five and twenty to . . .' are practically obsolete.

The work should not be taken in isolation, but should be linked with the interval timing already discussed in Activities 12 and 13. The children should be used to counting seconds in step with a large stop-clock; and they should, by now, be familiar with its minute hand which marks off a minute when the sweep second hand revolves once.

The children need to be told that although they cannot see the hour hand moving it is in fact moving slowly all the time: call their attention to it at, say, 15-minute intervals.

The old-style demonstration clock face with hands that could be moved independently is better banished from the classroom. An old clock or, better, a plastic clock face with a simple geared movement is essential equipment.

A useful set of activity cards can be made by writing, on the top half of the cards, sentences associated with a time. On the bottom half (using a rubber clock face stamp) give the appropriate time. Cut between the two along an irregular line.

The children then match up the cards.

The effect of the digital clock on teaching time is discussed further in Activity 41, which carries an important note on this.

40 *Towards a calendar: days and seasons*

By now the child is beginning to appreciate more clearly the idea of a year, helped by the recurrence of birthdays, Christmas, and school holidays. Children do, of course, speak of these things long before they can relate them to the seasons or to one another. The calendar, moreover, is not obviously cyclic to small children and we need preliminary work to help them grasp the concept of cyclic time changes over longer periods.

Make the class a circular chart of the week as a seven-day cycle, fitted with a pointer to help in counting on and counting back. A different colour for the two weekend days makes a good starting point. The usual British convention of taking Sunday as the first day is followed in the diagram.

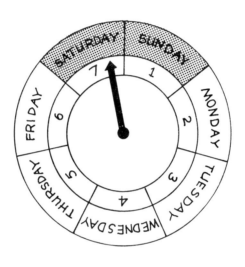

The children will need to see that this is not a device for marking the day and date, but a help in talking about days and weeks. With its help one can ask and answer many simple and useful questions. The work does, of course, relate to a child's knowledge of the multiples of seven, and reinforces this.

For example:

> Today is Tuesday. How many days until it is Wednesday week?
>
> How many days have gone by since last Thursday?
>
> Today is Thursday, and Christmas is in exactly four weeks.
>
> How many days is this?

Extend the work with a similar chart giving the months and seasons. The four quadrants in the centre can carry suitable pictures cut from cards or magazines and could be different in rural or urban locations.

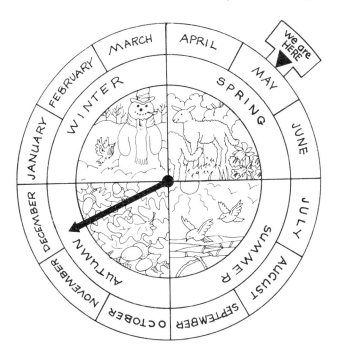

When not being used to answer questions involving the counting of months, the hand should be set to the approximate date angle so that the passage of the year is marked. Better still, another hand longer than the radius of the chart can be fastened at the back, carrying an arrow and the words 'we are here'.

41 Time in travel: the digital clock

Time is not an easy measure to teach because it is governed by so many conventions, all of which are found in ordinary use. Today all children will have met the digital clock and we need a device to ensure that they can read it. The 24-hour system will be deferred.

Make a digital clock display by making two parallel sets of four slits in a box about the size of a shoe box. Through the slits pass loops of paper on which the required digits are clearly marked: 1 on the first, 0–9 on the second and fourth, and 0–5 on the third.

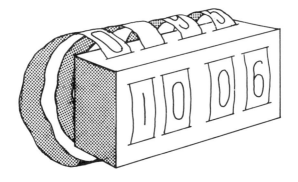

The display, which as described does not use the convention of an initial zero, can be set to any time for the children to read. They can be asked to transfer the time to clock diagrams made with a rubber stamp. A glance into the windows of any shop selling watches, or at the pages of any magazine or brochure advertising them, shows that digital display is becoming more and more common. In a few years Activity 41 may need to precede Activity 39, because battery-operated digital watches are then likely to be cheaper than those with ordinary movements.

Evaluation checklist (iii)

As we have presented it, Stage Three, the transitional stage that begins measurement using standard units, is quite short. Clearly, the activities involving time, collected together because they are still formative, are meant to be spread out over a longer period; but the point is now made that we use standard units in much the same way as we used the informal measures that did not involve standards of comparison.

The pupil will now need to know the units that have been introduced, together with one or two technical words such as 'digital' or 'graduated scale'. These words should appear in use and become part of the child's normal vocabulary, not special words used only in mathematics lessons.

For this reason, the sort of questions and activities put to the child are not once and for all test items. They are used only to make sure that there are no conceptual hang-ups, due perhaps to absence (or even the onset of a physical condition involving eyesight or hearing). Many children will be using the words freely in their conversation with one another or the teacher; but if there is any doubt specific efforts are needed. Examples could be

> Get the metre rod and measure the windowsill for me, will you, please?
>
> How many centimetres of chalk have I got left?
>
> Would you check whether this bottle holds a litre?
>
> Would you go into the hall and see what the time is by the school clock?
>
> (This question can easily be modified — e.g., 'see if it is eleven o'clock yet?')
>
> What is the number marked on the end of the scale?
>
> Find out how many 10-gram weights make the pointer point to 150.

A group of teachers might like to produce an extended list of such questions for their own use.

Primary activities — Stage Four

With Stage Four, measurement becomes a true application of mathematics. The three earlier stages, separated more for convenience in exposition rather than for classroom procedure, cover the premeasure activities. These are the concept of 'measuring' one quality such as length by an object used as a unit and the key concept of a standard measure. No number process other than counting was called for.

We are now going to relate measures such as centimetres and metres numerically. We shall begin to combine them by the processes of arithmetic and represent them on graphs. Purpose-made measuring apparatus such as tapes and trundle wheels can now come into use, although it is not until Stage Five that measure becomes fully operational. At Stage Four we are still measuring for the sake of measure: we are helping the children learn the techniques, and linking them up with developing number skills. Clearly, the teacher will want to seek every chance of making the transition from practising measure to *using* measure on tasks that call for it and cannot well be undertaken without it. Our final section on putting measures to use can be referred to from now on, and anything certain of being within the capabilities of the children can be tried out.

In the activities that follow we have sometimes chosen those whose results are of interest, scientific or technical, but at this stage it is the measurement that counts. A fully integrated treatment of a topic is possible only when the basic skills are properly developed.

The point is important for teaching and fits in here with our discussion of stages. The best place to learn to handle a vernier micrometer is in a workshop, on a job that needs to be measured using one; but this is not the situation in which one best receives instruction in the use of a metre/millimetre rule.

Nevertheless all our classification of activities is provisional and, subject to the development of concepts discussed, any appropriate work that the teacher can devise may be brought in or used as a substitute for anything described.

42 *Litres and millilitres: the fundamental submultiple*

The key idea behind the activity is the division of the litre into one thousand parts. This will probably be one of the children's first contacts with the thousandth as a number in use. At this stage the millilitre will not be used as such but only in multiples of 5 and more commonly of 10.

Refer to the earlier work on measuring in litres and introduce the symbol *l* (see page 12 for discussion of the standard symbols). From now on, instead of 5 litres, write 5*l*, although this is read as 'five litres' and not 'five L'.

Now introduce the 5-millilitre medicine spoon and a few medicine bottles with the standard 5-millilitre dose given on the label. Discuss the word 'millilitre'. It means that we divide up a litre into a thousand parts. Ask the children to imagine a small spoon that could be filled a thousand times with a litre measure of water, or better, let them handle a small mustard or saltspoon which will be of *about* the required size. This mustard spoon (or doll's teaspoon) holds about one millilitre, this medicine spoon holds five, this small bottle holds twenty-five.

Show that 'millilitre' is written m*l* and is so printed on most medicine bottles. Say that since the millilitre is small we do not often use it singly, but in fives, tens, or even fifties. Let the class, orally, build up a hundred in fives or tens, and a thousand in fifties.

Have available some plastic beakers clearly marked, and a few plastic pipettes in 10 m*l*, 20 m*l*, or 25 m*l* sizes. The pipettes, of course, will be used with water only, and the children need to be shown how to draw up the water to just above the level with a finger near the top, making the final adjustment to level by allowing air to leak in. They will not find it easy at first.

Discuss simple numerical relations. The half litre already met is now 500 m*l*, the quarter litre is 250 m*l*. Let the child see that a litre bottle will fill a 250 m*l* beaker four times.

Note that the child is *not* 'proving experimentally' that 1000 m   make a litre. The relations between measure are a matter of definition, not demonstration. The pupil can show that four 250 m   containers do in fact fill up a litre, but the aim is to familiarize only, and to practise the skill of transferring measured quantities of water from one container to another.

Make a collection of labelled bottles whose contents are given. Examine the bottles before including them. Perfumes, medicines, and the like will be in millilitres, but beverages and household fluids of British origin will usually have their contents in fluid ounces as well. Tell the children that this is an older measure that is beginning to go out of use.

At this stage, for dealing with capacity, we do not recommend introducing the measuring cylinder graduated in millilitre or five-millilitre divisions. This is difficult to read accurately. The idea is simply to establish the useful smaller measure as multiples of the millilitre, itself seen as a small fraction, a thousandth part of the litre. That the work helps to form the child's concept of 'a thousandth part' is important, and is fundamental to the interaction of measure and mathematics in the teaching process. We do not want people to say that a group of children is not yet ready to deal with metric measures 'because they can't do thousandths'. The point will arise again with decimal fractions in number work.

Extend the work by bringing in wine or other bottles marked in centilitres. This measure is very common outside the UK and can be discussed as a practical unit, one hundredth of a litre and hence ten millilitres. See note on page 12.

43   The centimetre scale: starting from zero

This is given as a separate activity which moves on from Activity 34. The scale used does not match up a length against numbered centimetres but against a length divided into centimetre intervals, numbered at the end of each. The difference between the two is of more importance to the teacher than the child, since it is leading eventually to the concept of continuity in measure. At first, in class, it is simply another way of labelling the centimetre units.

Repeat and extend the work of Activity 34 using 15 cm rules graduated in centimetres with the centimetre marks numbered from 1 onwards, as in the diagram below.

They can be made in the same way, using graph paper, as the earlier rules. Have the numerals rather smaller and with the first digit to the right of each division so that the zero can be put in. Discuss the use of the zero, which marks the starting point at which we have not yet measured off any centimetres at all.

We have now measured lengths in three stages:

1 By counting the centimetres that fill an interval.

2 By numbering the centimetres to save counting: we merely record the last number.

3 By numbering each centimetre interval on the dividing line, so that we count it, beginning from zero, only after covering the space it occupies.

The first rule corresponds to the line of centimetre cubes used earlier. With the second, one can easily stop halfway between centimetre three and centimetre four, and by careful preparation of extra duplicated measure sheets the teacher can provide lines that require half-centimetre intervals to be estimated. The work would be recorded as before, using the $\frac{1}{2}$ symbol.

44 Units and submultiples: linking larger and smaller

It is now time to make a link between the centimetre and the metre, which have been introduced so far as if they were independent units, and to link these in turn to litres and millilitres. The work collected here is presented as a single activity for convenience of reference. In fact it must be spread over many weeks, and it uses apparatus that should be common classroom equipment. Here we begin to formalize the measures children have been learning to use incidentally.

Educational suppliers now offer metre rules divided into centimetres. Not long ago only the standard laboratory metre rule graduated in centimetres/millimetres was available. The millimetre is too small at first as a classroom unit, although later in the workshop it comes into its own in that it allows most measures to be given as whole numbers.

Let the children compare their centimetre rules with the metre rods graduated in centimetres and see that the units are the same. Refer now to the name centimetre and give its meaning as one hundredth part of a metre. The word can now be related to the centilitre discussed in Activity 42, which is one hundredth part of a litre.

Collect together the words

litre   metre
centilitre   centimetre
millilitre   millimetre

It would be worth while asking the children to guess at the last word, which has not yet appeared in an activity, although some of them are likely to know it anyway. Now show them and let them handle a millimetre scale and point out that the unit is rather small but that they will be using it later.

Extend the discussion as far as convenient, mentioning the French or Latin for one hundred and one thousand. The children need to feel at home with the prefixes because we use them in many other measures.

Make a complete list of the symbols

l   m
c*l*   cm
m*l*   mm

referring to page 11 for details of correct usage. Children learning the older imperial measures did so against a background of general, if erratic, social use. They are now being taught what, for many older parents who have had no contact with science or modern technology, is an alien system of units. It follows that every opportunity should be taken to get them familiar with the measures through handling classroom equipment, so that the naming and listing of units brings together what is already known informally. It is advisable to stick carefully to the internationally agreed notation now found in scientific and technical works and in the usual examination syllabuses. Such forms as 'gms', 'cms', or 'ccs' are to be regarded as obsolete, current in the UK only before international agreement.

Children will now be able to return longer lengths in centimetres. Note, however, the conventions. A table top is not 1 m 83 cm, although it can be measured by a child at first as 1 m and 83 cm extra. The two are normally put together directly as 183 cm. This is standard usage, and is obviously quicker and more easily recorded. (It resembles the milkman's use of 11 pints instead of 1 gallon 1 quart 1 pint – a form that only survived in school textbooks.) A longer distance, such as the length of the corridor, would be given to the nearest half or quarter metre: to give it in centimetres is to impose a false appearance of accuracy.

We should remember, though, that we are helping children to learn mathematics, not training them as engineers. The form 1 m 83 cm can be used whenever it is convenient, and so can the alternative notation of 1.83 m. This, with the decimal point thought of as separating the fractional units from the initial metre (or indicating the start of the tenths) is one of the practical approaches to the decimal system. Whether this form is or is not to be used in a building specification is not relevant at this stage. We may, therefore, for educational rather than technical reasons, want to get children to record subunits using the decimal notation.

45 *Linear scales: representation in measure*

The use of a scale, in which one measure represents another, is a fundamental mathematical device. It is not, in itself, a form of measurement but a technique for handling both measures and pure numbers in diagrams and graphs. This activity suggests an approach that should be used whenever opportunity offers, and is not intended to be worked through in isolation.

The concept of scale is best developed informally by using the word in context. Many children are familiar with scale models. Sometimes, as with model cars and trucks, they are sold as such so that children will themselves use the word.

Questioning can often help, using models from the class collection.

Is this a scale model?

Would you say this was a large-scale or a small-scale model?

Do you know what the scale is?

Such questions can well show that the child has grasped the basic concept in out-of-school activities with toys. A child can know, informally, the meaning of a scale of 1/20, long before meeting classroom exercises in fractions or scale drawing.

Other toys, like dolls' houses, can be thought of for teaching purposes as scale models, but are not normally associated, either by suppliers or purchasers, with the numerical concept.

The example of the toy car suggests that the first ideas of scale can precede actual measurement. The real car is fifty times as long as the model. If we put fifty of the models end to end they would stretch from the front of the real car to the back. The small car is then being used as a unit to measure out the large car: it differs from the work of Activity 28 only in that the unit object is the same shape as the measured object, and when two things are exactly the same shape but of different sizes one can be thought of as a scaled-up or scaled-down version of the other. In this way the word 'scale' can become part of the child's language.

The concept now needs to be linked to actual measures, but need not be elaborated. It is a topic that will be met and used frequently after the primary school, where it will accompany greater skill in making accurate drawings and measurements.

Children can be asked to measure tables or classrooms, or to bring from home the measures of rooms or gardens if the teacher knows this is possible. Scale drawings can then be made in two stages: a rough freehand dimensioned sketch which shows what is wanted and then a final diagram drawn with a ruler to a simple scale. The scale can be written out in full, as, for example '2 cm represent 1 m').

The work is best kept simple, and should avoid being a premature attempt at technical drawing. We do not recommend a formal treatment of map scales at this stage, since this work is very commonly done in the first year of secondary schools, although a later activity (56) suggests an extension.

46   *Multiple units: grams and kilograms*

When the metric system was first devised, the gram was intended to be the standard mass and the kilogram a multiple unit. Under the SI system it is the kilogram that is standardized, but the older names are retained. It is hardly necessary to make this point to children: it is enough that they should be able to relate gram and kilogram as metre and kilometre.

Weighing with standard units was introduced (Activity 35) using 10-gram masses. Simply by holding a few of these plastic discs in one hand and a bag of potatoes in the other, the child should realize that a heavier unit is needed for many purposes.

Let the children handle the kilogram. It is a good idea to have a kilogram attached to a cord running over a small pulley, so that a child can pull down with the same force that is needed to pick the mass up. They can be told that it gets its name because it is 1000 grams and be given the symbols, linking now with metre and kilometre, already met as a distance in Activity 33.

kg   g

km   m

The kilometre is referred to again in Activity 43.

If the class has a compression balance whose pointer swings from zero to 1000 g, let the children put masses on it to satisfy themselves that the kilogram is in fact formally equivalent. If masses of 500 g or 250 g are available (a complete range is obviously worth trying to get), show that 500 g of sand plus a 500 g mass give the same reading as the kilogram or that they balance on an ordinary pair of scales. The classroom should then have available a 'personal' bathroom scale calibrated in kilograms, so that children can find and record personal weights each term.

Collect, if possible from the children's homes, examples of one kilogram of commodities such as potatoes, coal, sugar, and the like to put on temporary display. The class need to handle these to get the feel of their weight, in case they do not get the opportunity at home.

47 Heights and distances: straight lines and curves

Subject always to the condition that children do not repeat activities from year to year without good reason, there are numerous ways in which they can practise linear measure and extend their conceptual grasp of the process, linking it, for example, with growth in time, the tracing of curved paths, or graphical representation. The examples which follow show the sort of measurement activity that can be introduced and call for techniques that suit the circumstances.

By now it should be possible to write up some activities as work cards. We shall give them as suggestions which can be interpreted as instructions to children in the light of the teacher's experience. The suggestions are numbered.

1 Stand one child on a mark in the middle of the playground holding a cord a few metres long. Another child walks round the first with a metre trundle wheel, keeping the cord taut, starting on a line drawn out from the first mark, or marked with another cord or rope. If the distance between the children is measured with a metric tape to include the length of their arms, one can investigate the result of doubling the distance between them. Does it appear to double the trundle wheel reading? It is better not to go beyond this point at this stage, since the measures as obtained are hardly accurate enough to permit conclusions about the properties of circles.

2   Most primary classes, of whatever year, could have a means of finding the height of pupils. Vertical scales two-metres long and printed on strips of paper are available, or a metre rod calibrated in centimetres can be set up on the wall, with its zero one metre from the ground. This requires the child to add 100 cm to each reading, and is probably a better teaching aid than a purpose-made scale. A child who had not done it before would probably need to be shown how to hold a flat ruler horizontally over the head of the one being measured. The results can be recorded in centimetres on a table or transferred by measurement to lengths cut from a roll of paper tape, which are then marked with the child's height and name. The strips can be pasted on to a large sheet of paper to give a graphical representation of the heights of the children. If the work is done as a class project the teacher could write the heights obtained on the blackboard and discuss arranging them in order of height. (See also Activity 31.)

If this exercise is done at the beginning of the school year it can be repeated at some later time, and the difference made up with an extension of coloured paper, so that both height and growth appear.

3   Supply a large ball of thick wool or yarn. In rural districts the remainder of a ball of baling twine can usually be acquired. Discuss a way of measuring its length without pulling it all out into one long line. One could, for example, measure a length marked by two lines across a corridor, unrolling and rolling up again a section at a time. Questions to children:

What would be the best distance to mark out between the lines?

How can we check that we don't miss bits of the twine or count them twice?

Can you think of a way of finding the length without measuring the lot?

Would weighing help?

There are two possible ways of finding the total length by weighing. One way is to measure out a suitable length such as 10 m whose weight is then compared with the weight of the ball, the other is to weigh out 10 g or 100 g of the wool or twine and measure its length. The exercise is useful in that one can discuss the idea of a reasonable approximation. It might be good tactics to adjust the weight of the ball before giving it to the children. This can avoid awkward arithmetic.

4   By using a trundle wheel or a metric tape place markers in a playground or playing field to form a square of side 25 m. The object is to get a closed route of 100 m, and variations on the square can be discussed with the children. Get someone to check, with the wheel, that the total distance is near enough to 100 m, adjusting one of the markers accordingly if it is not. By walking ten times round the square, taking a marble from a box of ten at one corner each time round to keep the tally, a child can walk one kilometre. It is true, of course, that a measured distance thus covered seems much longer than the same distance covered during a walk. The children can be sent out in pairs during other lessons or activities. They can take a stop-clock with them or calculate the minutes in any other way that has emerged from their work on time. The numerical connection between metre and kilometre has already been seen in the previous activity.

48   Estimation of area: counting the bits

This activity brings in a technique for estimation of the area of an irregular shape by counting squares, and in so doing introduces the centimetre square as a unit.

Discuss the function of leaves on a tree, how they absorb sunlight and 'breathe', so that by being thin they have the advantage of a large area for a given weight. If the children collect some leaves of irregular shape, such as ivy or sycamore, we can discuss the amount of leaf (or area of leaf) exposed to the sun and air. Direct the discussion towards the use of squared paper and get the leaves traced on to centimetre graph paper. Each square is the same size as the face of a centimetre cube as found in the arithmetic rods. The children will have covered areas with square units (Activity 37) but we are now going to use centimetre squares.

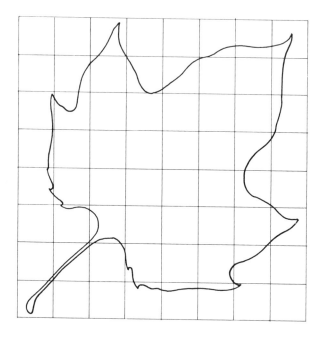

Discuss the problem of counting the squares, of keeping count of the **whole squares and allowing for all the bits round the edge of the** outline. It is usually possible to make progress with such a discussion by asking leading questions.

A reasonable technique is to count as whole squares those with half or more inside the trace, and to ignore those with more than half outside, marking the ones to be counted, as the whole squares can be, with a dot. The areas in centimetre squares can then be compared directly. Note that the term 'centimetre square' means a square of side one centimetre. Its area is one square centimetre or one centimetre squared, but this latter measure applies to areas without reference to their shape.

One can also obtain transparent plastic grid sheets marked with squares which can be put over an outline. A felt-tip pen with water-based ink can be used to mark the squares temporarily to help count them. We suggest that this apparatus, since it does not produce a permanent record, should be used only after the principles are clear.

49 Calculating areas: a special shape

The result that follows was once the usual starting point for area, the product of the measures of length and breadth (or 'length times breadth'). Although the result *only* applies to rectangles, whose configuration allows the area to be calculated instead of counted in squares, the process is of importance and needs to be thoroughly understood. We do suggest that it is not introduced until the earlier direct measures of area are quite familiar.

It is probably worth while to repeat Activity 48 in a prepared form, by giving the children a supply of square units and some rectangular cards of an exact number of units in length and breadth. The children by this stage should be able to put down the squares neatly without gaps or overlap, giving them an arrangement ready to be discussed.

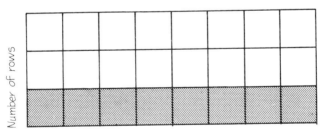

Refer to Activity 48 and ask if the rectangles are easier to deal with. Obviously yes, since the squares are arranged in neat rows with no odd bits that do fit the outline. Ask if it is necessary to count all the squares, and eventually draw out the suggestion that one row is enough if we count the rows beginning with a small rectangle where this is easy to see. This is a very big step forward in mathematical thinking, and needs to be taken slowly, probably with some repetition over a period of time. Certainly children should have in front of them the covered rectangles as the discussion begins. This is not an activity to be begun and completed as a single lesson.

We suggest that the line of thought is written up in full.

Number of unit squares in row	$= 9$
Number of rows	$= 5$
Total number of unit squares	$= 9 \times 5$
Total area	$= 45$ unit squares

The class or group can be asked to do several simple examples orally. Spend some time talking round the result. Ask numerous questions such as:

Is the number of unit squares in the row the same as the length in units?

Can you explain why?

How did you know how many rows there were?

How many rows would there be if the breadth of the rectangle were 8 units?

Can you tell me, without a drawing, how many centimetre squares there would be in a rectangle 10 cm long and 6 cm wide?

What happens if the breadth is the same as the length?

The last question, of course, is important. It introduces an entirely new use of a word already familiar in the context of a new notational concept. The question is best delayed until the children are quite sure about the area of the rectangle. Here we multiply the number by itself and it gives us the area of a square whose side in centimetres or other units is the number concerned. Later it can be explained that this is so often used that when we multiply any number by itself we say we are 'squaring it' or 'finding its square'. The square of 3 is 9. We sometimes say 3 squared is 9 and write $3^2 = 9$.

Extend this work by bringing in easy mental examples in number only, relating to the dimensions of squares. The children have now to make a quite different use of the word 'square' and will probably need plenty of examples, both mental and written. It is important to reinforce this key piece of work. By now a more straightforward approach is possible, and we suggest issuing work sheets having suitable rectangles, either divided into unit squares or drawn of integral dimensions so that the children can measure them.

There is a point for the teacher to note. It is often argued that one can only multiply numbers, not centimetres or any other measures. The area is derived from the number of centimetres in length and in breadth. It is therefore held to be incorrect to write:

length	=	10 cm
breadth	=	5 cm
area	=	10 cm × 5 cm
	=	50 cm²

A correct version would then be claimed to be:

length	=	10 cm
breadth	=	5 cm
area	=	(10 × 5) cm²
	=	50 cm²

The argument may be mathematically sound, but it is rather pedantic in view of the technical use made of the first version. It is better to think of the usage as subject to two distinct conventions. There is no reason why both should not be used later, but the first version is the one the pupils will use if they do model making, carpentry, or the like.

Moreover, the symbol cm² which this work introduces to the children seems to be constructed as cm × cm using the index notation. The argument should be quietly dropped.

150   *The ever-rolling stream: epoch time*

We have, so far, dealt with time as an interval between events and time as marked by a cycle of recurrence. The concept of historical time, of the gap that separates us from the past, is not readily grasped by children. We often speak of a time scale, but our own adult ideas of time past are not very precisely held. The concept of a time scale cannot be taught, but there are activities which help us to establish it.

One way is to make a year indicator from long strips of 2 cm graph paper, three squares wide and 365 squares long. Get a child to count off the 31 days of January and mark this point. Let other children deal with the other months of the year, eventually entering the names of the months along the top row of squares. Then, by reference to a calendar, get the days of the months filled in by the children. 1–31 for January, 1–28 for February, and so on. Finally, along the third row of squares enter or get a group of children to enter the numbers 1–365.

The class now has a linear calendar which can be displayed on the wall. Each morning stick a coloured square over the previous day in the bottom row, filling in also any squares corresponding to holidays and weekends. It is better for two classes to cooperate on this, each making a linear calendar, so that children who begin one of them in January can reach December on the other, even when moved up to the next class.

Since the linear calender on the scale suggested is over 7 m long, the work can be extended by discussion about the number of strips that would go all round the room, or all round the school hall or playground. This can give an idea of a life span in relation to a child's own age.

The very early work in observing the growth of plants can be extended to a discussion of the life cycles of plants and animals: emphasis on the time spans of the cycles helps to get the children thinking in quantitative terms about duration.

It is worth noting here that most children's literature is more or less static in time. An old man in a story has always been an old man, and children are rarely followed as they grow up. This is what children expect of a story and there is nothing wrong with this; but it does mean that the concept of the passage of time is not often found in a child's general reading. It is history teaching that suffers most from the lack of an effective time scale, but it is in measure and mathematics that most can be done to establish the concept. History puts it to use.

51   *The personal record: a general exercise*

Several of the earlier tasks have required children to weigh or measure themselves or one another, using the results in various ways. Children can also keep a personal record.

If a personal record book can be made so much the better. The idea of this 'Book of Myself' or 'All about Me' is to collect and record any details that children or their friends might find interesting. The book is best kept with each page dated. Some of the details would doubtless be irrelevant to measure and the use of number, but children can be encouraged to include as much as they can. There could be an emphasis here on neatness at this stage. It is a good occasional exercise to set out information, numerical or otherwise, in a clear and balanced format. Typical entries are:

My name is ——————————

My address is ——————————

My age is ——————————

I am ——————————cm tall

I weigh ——————————kg

My chest measures——————————cm

The area of my footprint is ——————————cm²

I can walk a kilometre in——————————minutes

My shoe size is ——————————

If the children can be allowed to take the books home, many parents will no doubt cooperate by helping children to extend entries. The children should write their own entries as complete sentences. The books are a useful source of data for graphs and other exercises.

52 *The square metre: using a quadrat*

The quadrat is a square frame often used to mark out areas of ground for biological sampling. We shall use it to familiarize the children with the larger unit of square measure, the square metre or metre squared.

Cut four dowel rods to one-metre lengths and make them into a square frame by drilling holes in two adjoining sides of four small cubes, glueing the dowels into place. Let the children examine the apparatus and realize that it is used to mark out a metre square, among other things for counting weeds in fields or lawns.

Children can try this in pairs if a school field or suitable area is available. Do not worry about correct statistical sampling: merely drop the quadrat in a likely area where dandelions or plaintains seem to grow in reasonable counting numbers. The dandelion heads can be counted directly, the plaintains by putting a small counter or light coloured bead on each. Then record the number in one square metre of the grass.

Extend this work at a later date by asking how many children can stand comfortably in one square metre, and then consider its area counted out in square centimetres. Ask how many centimetres make up the length and the breadth, referring back to the earlier activity on the area of rectangles and squares. Consider the product 100 × 100 as the number of centimetre squares equivalent to the quadrat.

Establish the symbolic statement

$$1 \text{ m}^2 = 10\ 000 \text{ cm}^2$$

By joining large sheets of centimetre grid graph paper, make a wall display of 1 m² of graph paper, and by discussion, let the children arrive at the conclusion that it contains 10 000 small squares. It is useful for them to see such large numbers displayed. We can see ten thousand squares!

By referring to the earlier Activity 49 with rectangles, it should now be possible to extend the work to the areas of rectangular rooms or spaces given in metres. Perhaps the children could draw a plan to scale, and discuss whether the steps that allowed us to calculate the area of the rectangle in square centimetres can be used for rectangles measured in metres.

Use the same format:

length of room	= 11 m
width of room	= 7 m
area of room	= (11 × 7) m²
	= 77 m²

Evaluation checklist (iv)

We have now reached the point where, at the end of Stage Four, the children should be capable of direct measurement of heights, lengths, weights in the correct standard units, using halves, quarters, or the decimal notation when appropriate. They should be able to handle these measures with straightforward arithmetic processes as discussed in *The Third R*. They are now using measure correctly, with both decimal fractions and the submultiple units that avoid them. The measuring situations have been simplified to avoid the difficult concept of accuracy to a given number of digits. The children are for the most part at the stage where, given the experience in application that can only come in the context of real life activities, they should be able to cope with measure at least in the home.

Their working vocabulary should now include words such as:

accurate	represent	flexible
approximate	diameter	rigid
estimate	circumference	irregular

which will have been used by the teacher in discussing the activities so far undertaken. They will also be quite familiar with the following standard units, and have personal criteria for estimating them:

centimetre	gram	millilitre
metre	kilogram	metre squared
kilometre	litre	centimetre squared

This familiarity should extend to the confident use of the appropriate apparatus.

It is usually practicable to reserve sets of the articles suggested for measurement exercises to use for evaluative tests. They should cover the required range of dimensions and be marked in some way so that the results can be checked. For any one child evaluation can take place at any time, and need not be done if the evidence already available is good enough.

Examples of test activities:

Measure the length of this piece of cord.

The cord should be coiled up, and between one and two metres long; the child would perhaps stretch it out on a table and use a metre rule, returning the result in centimetres. The child might instead use a tape: the point is that the length, correct to one centimetre, is given in centimetres.

Weigh each of these blocks on the pointer scale, in grams.

Now work out the total weight of the blocks.

Using your 30 cm rule, measure the lengths of the lines A, B, C, D, E on this test card, guessing the half centimetres.

The card is prepared as for Activities 34 and 43, but is reserved for test purposes.

Measure the side of this square and also the distance diagonally across it.

A 10 cm square will have a diagonal of 14.14 cm which the child will probably return as 14 cm.

The next question could be:

Is it exactly 14, or just over or just under?

A coil of wire 100 m long weighs 500 grams. Work out the weight of a kilometre of the wire.

Use the model of the digital clock in Activity 41 and set the loops so that the digits show the following times: a quarter past nine, a quarter to twelve, half past one, twenty to three, midday. . . .

The questions can easily be devised to correspond to the level of ability expected of the child.

Tell me the areas of these two rectangles and the square.

Measure up the plain rectangles and find their areas.

The first set would need to be cut from centimetre squared paper, the second from plain card. Both would be cut to a whole number of centimetres.

Use the 20 ml plastic pipette to fill this bottle up to the mark on the side, and then tell me how much water is in it?

The mark should be at 50 ml or 80 ml, carefully checked in advance: the 50 ml is the more difficult of the two. A pupil who can approach these and similar tasks with confidence, who has passed through all three earlier stages, can be taken as proficient in handling measure to the required level, and as having a sound conceptual foundation on which further skills and understanding can be based.

Primary activities – Stage Five and the final objectives

This is the last of our rather arbitrary stages, and should if possible be reached during the last primary year. The actual work of the stage in any one measure can begin whenever a teacher feels it to be possible. The activities now take children to the point at which they should cope with the general uses of measure. We have already pointed out that for most 'domestic' uses the work of Stage Four, extended for the adult by the practical skills in cooking, home making, or D.I.Y. that provide measure with a context, is sufficient. Beyond Stage Five the work becomes specific: the pupil may learn to use a chemical balance or a stage micrometer for a microscope, may learn to work metal to tolerances beyond the scope of the household odd-jobber, may learn to make complex indirect measurements such as electrochemical equivalents.

But all these skills, which need to be taught in the context that requires them, depend on a sound grasp of the measures listed on page 11 and the ability to combine them by the ordinary rules of arithmetic. Some of the activities given here, since they provide a context for measure, begin to bridge the gap between primary and secondary education. It is still true, however, that they are done by the pupils as an exercise in measure, and not as scientific experiments that need measure to arrive at the required conclusions. By the end of the stage, however, we hope that the children will be able to use measure in the context of primary science.

It is now appropriate to state what we should expect a child at the top of a primary school to be able to do with measure. These are the so-called 'educational objectives' to which the activities in the book have been leading and which will be completed by Stage Five.

May we, however, repeat the warning given with a similar list in *The Third R*. These educational objectives are not test items to be checked over and then forgotten, like O-level passes in the incidental subjects required for entry into technical or other courses. The recorded fact that a child of ten successfully calculated the areas of rectangles does not help anyone if the sixteen-year-old seeker after an apprenticeship then gets hopelessly muddled on a similar task.

These skills have to be maintained by being kept in use. Ideally, secondary schools would have a programme of some sort of quantitative study that keeps measure and calculation with measure alive throughout an entire school career. Such a programme, based perhaps on ecology, archaeology, or aspects of geographical studies, could be of great social value. Obviously scientific, mathematical, or workshop studies do this anyway, but some pupils who channel their efforts into the arts or humanities may be allowed to regress to incompetence in handling number and measure.

In the primary schools we should now organize activities that keep the whole set of objectives in review, including computation. This list of Stage Five objectives, as given, implies all the earlier skills that need not be listed separately; but we regard every item as essential. No children should leave primary school without every effort having been made to get them to the level of competence set out below. The paragraphs are numbered for reference.

Length

1 Children should be able, using rules or tapes, to measure objects, intervals, and distances in any of the forms in which such measures occur in daily life, in metres, centimetres, or millimetres, combining these units with the decimal notation if needed.

2 They should be able to perform such arithmetical operations on these measures as can be shown to be needed in general situations; but in addition they should have sufficient grasp of the principles involved to be able, in their secondary schools, to acquire the special skills needed in laboratories or workshops without further *basic* instruction.

3 They should be able to operate with the kilometre as a unit of distance and work with distance, time, and speed in straightforward practical calculations.

4 They should be introduced to the linear scale representation of lengths and distances as they appear on graphs and diagrams, and be able to construct scales in simple examples.

Capacity

5   The primary pupils at eleven should be able to work with litres and millilitres using pipettes and graduated cylinders. They should be aware of the litre as equivalent to 1 dm³ and of the very important fundamental result that a millilitre of water is equivalent to 1 g. (See Density, below.)

Area

6   The concept of the centimetre squared and metre squared as units of area for both irregular and rectangular shapes should be well established and the children should have techniques for measuring area of plane shapes in general and for calculating the areas of rectangles. The difference between area and perimeter must be clearly understood.

Volume

7   As for area, using the centimetre cubed and metre cubed, with the relation between the litre as a decimetre cubed and the millilitre as a centimetre cubed. Calculations limited at this stage to cuboids.

Time

8   All children should be able to read and record the time in hours and minutes using the usual clock notation, the digital clock, and the 24-hour digital clock. Although this introduces skills involving tabulated information, they should be able to read timetables in the 24-hour system and convert from and into the 12-hour system. They should be able to calculate the interval between two given times, and should be able to measure and record lapsed times in seconds.

Mass

9   The children should be able to weigh objects in balance scales or on kitchen or domestic weighing machines in general, to the degree of accuracy required in use.

Density

10   They should have an informal concept of density, realizing that some substances are heavy for their volume and others light. They should be able to use the equivalence of 1 litre (or 1 dm²) and 1 kilogram of water as a convenient standard, and relate this to flotation.

Temperature

11 Children should be able to read thermometers calibrated in whole degrees Celsius, and use them in very simple experiments such as constructing a cooling curve. They should be aware of the use of negative numbers to represent temperatures below zero.

Angular measure

12 Children should be able to use the right angle and its easy fractions as a unit of measure, and relate these to the eight principal compass points. They should be able to use a 360° protractor to an accuracy of about 2°, and be able to express the right angle and its easy fractions in terms of degrees.

13 Pupils should be able to work with angular velocity as revolutions per minute for complete revolutions only.

General

14 It is expected that work in measure would link up with work in geometry. See *Children Learning Geometry*. (This book deals with geometry for the age group 5–9 in an entirely nonquantitative setting, but the appendix suggests a more formal treatment for older children.)

15 Any reference to computation throughout this book assumes that the children have the necessary number skills when they are called for. A very detailed discussion of these is given in *The Third R*. A primary pupil who can tackle most of the activities that follow should have no difficulty in meeting the final stage objectives set out above, and will also be gaining experience in both mathematical and scientific thinking as measures are put to use. Teachers or mathematics consultants in primary schools who refer to this book in planning lessons might care to discuss the objectives with the secondary schools to which their pupils pass, so that those who take over at least know what has been attempted and what those of their intake who are not disadvantaged should be able to do. There are, deliberately, no evaluative procedures set out after Stage Five. Instead, the teacher is referred to the section on children using measure, which lists a number of practical situations that demonstrate a child's grasp of measure in use.

53   *Millimetres: the need for accuracy*

Up to this point lengths have only been measured to the nearest half centimetre. Young children can take readings in millimetres if shown how to do so, but they are rarely accurate. It is much better to wait till the stage when the children have more number skills, and can readily link up the millimetre with the centimetre as a decimal of it, as well as a submultiple of a metre.

Strips of centimetre/millimetre graph paper can be used to help discuss the division of the centimetre unit into ten parts. Each part is called a millimetre, a word already suggested in Activity 41. Questions could be:

How many centimetres in a metre?

How many millimetres in a centimetre?

How many millimetres in a metre?

This is an opportunity to discuss again the prefix milli- and relate it to the millilitre, asking the children a possible meaning for millisecond and milligram. Activity 42 has mentioned this prefix, but only in passing.

The children can cut a strip just over 15 cm long and about $1\frac{1}{2}$ cm wide, marking it out in cm from 0 to 15. It should be as in the diagram, and can be stuck on a strip of thin card if required – although the pupils will go almost at once to a proper millimetre rule and only need the strip for discussion and a preliminary exercise.

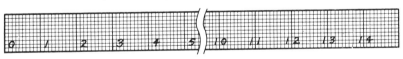

Use a measuring card of labelled lines as for Activities 34 and 43, but this time measure in centimetres and millimetres, recording the results as centimetres to one place of decimals. This stage is very important, and could link up with the early lessons in the decimal notation (of which this is a useful application). The work should in any case be coordinated with the school policy on teaching decimals.

After completing these measurements, ask pupils what the results would be if the readings had been taken in millimetres. The formal SI requirement that detail measures should be in millimetres should be relaxed in a primary school. We actually want the decimal points that the official form avoids!

Now issue rules, graduated in centimetres/millimetres only, and repeat the work with other duplicated cards and objects such as matchboxes, recording both in millimetres and centimetres with a decimal point. The best type of rule is of transparent plastic that does not cover up the object or diagram measured. Some plastic splinters dangerously if mishandled by bending, but safety grades of school rules are available.

Extend the work by measuring accurately cut pieces of lath or dowelling of less than one metre, but do note that unless these are carefully and squarely cut and smoothly finished, the measures taken along the various side and edges may differ by as much as a millimetre. Express these both as millimetres and as centimetres with a decimal. Although this exercise introduces and practises the use of the millimetre, the unit is too small and precise for most classroom measures. These can continue to be in centimetres. It is not advisable to get children to measure dimensions greater than a metre in millimetres. To do so implies that they can work to four figure accuracy and they cannot do this. An overall accuracy of about 1:100 is all that can be expected of most children under ordinary class conditions, and activities should, in general, be planned with these limits in mind.

Teachers will note that one likely source of inaccuracy in using a graduated rule is the parallax error that arises if the eye is not immediately over the end of the line being measured (see diagram).

Parallax error

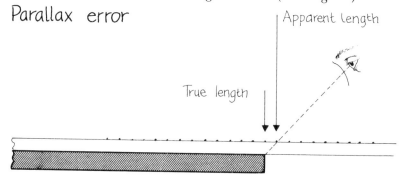

This point is not really suitable for discussing in a primary class, and probably the best compromise is merely to tell the children to look straight down on the ruler.

54 Areas of rectangles: the millimetre squared

The work on the millimetre as a unit of length can now, if the teacher wishes, be linked up with the millimetre squared as a unit of area, repeating Activity 49 in a modified form. Whether or not this work will be worth doing depends very much on the teacher's judgement. As a practical activity the errors in marking out rectangles accurately to millimetres with pencil on paper, even squared paper, will be too large to justify the calculations, but the work could be worth doing as a numerical exercise where the diagram serves to give meaning to the results obtained.

Issue pieces of centimetre/millimetre graph paper (whose main use in a primary school will be restricted to such purposes). Ask the children to try to mark out a rectangle, say, 8.3 cm by 4.1 cm, giving as an alternative expression 83 mm by 41 mm. The dimensions are supposed to be exact, but the diagram as drawn is not likely to be accurate even to the nearest millimetre. Now refer to the row count of Activity 49 using the centimetre squared, and try to get the pupils to transfer to a count in millimetres. They will not, of course, count the individual squares, but use the relation 1 cm = 10 mm. Hence deduce the area of the rectangle in millimetre squared:

length of rectangle	=	83 mm
width of rectangle	=	41 mm
area of rectangle	=	(83×41) mm^2
	=	3403 mm^2

This is a spurious accuracy in terms of the diagrams as drawn, and in practice the area would be returned to 3400 mm^2, but the reason for this takes us too far too fast. The exercise is intended solely to apply the principle of the row count to the smaller unit, for the benefit of children felt capable of extending the concept. Since the activity can be seen, at least in school, either as calling for the decimal multiplication 8.3 $\times$ 4.1 or the submultiple calculation 83 $\times$ 41, it could also be of value in work with decimals.

The work can be extended by noting the equivalence 1 cm² = 100 mm², and, if the children continue to follow, the equivalence 1 m² = 1 000 000 mm². Using large A1 sheets of cm/mm graph paper it is worth while making up a wall display of 1 m² of cm/mm paper, which now contains one million small squares. Children like to say that they have *seen* a million. This graph paper is expensive, but the display can always be stored for later use.

55   Linear dimensions: instruments of measure

The basic tool is the graduated rule, but there are many instruments designed to measure intervals where a rule is not easily used. Large plastic models of these precision instruments are now available for primary schools. They are not essential to our Stage Five, but give useful supplementary practice and insight into measure.

It is always difficult to establish priorities in spending limited funds, and general classroom supplies such as paper and felt-tip pens are probably more important than purpose-made instruments. One would want to select after much deliberation. Possible items, all to be found in the usual educational supply catalogues, are:

Inside and outside calipers

Caliper gauge

Feeler gauge

Depth gauge

Pedometer

Taper gauge

Micrometer screw gauge

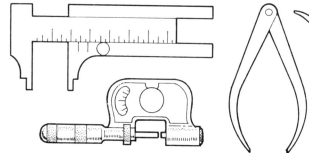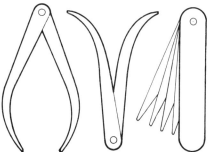

A school with an active PTA could probably set up a 'measure exhibition' and borrow actual instruments for a few days: these could be compared with the plastic versions if both are available. Teachers should, by using them, make quite sure that they can deal with any queries that are likely to arise, but by and large the instruments are available for interested pupils to experiment with. We do not suggest formal lessons on their use. It is fairly easy to devise leading questions that help the child to establish the use of an item, and in this way retain the knowledge more effectively than if merely given a demonstration.

Example:

Pupil: What is this?

Teacher: A taper gauge.

Pupil: What do you use it for?

Teacher: Bring over those two small pieces of plastic tube. Put the point of the gauge in one of them and tell me what mark it comes up to. Now put it in the other. Why does it show a different mark? If we used a wider piece of tube what sort of mark would be shown?
What do you think the marks measure?

Teachers will know that it is a demanding process to teach children by *not* answering all questions directly, replying instead with a question that leads the child's mind in the direction of the required answer. It cannot be done all the time, but it is worth trying when the opportunity offers: it is easier to remember something worked out for oneself.

It is clear that the class will need a collection of items, such as the pieces of pipe or conduit mentioned, suitable for measurement by the devices on display.

56   Distances on maps: the use of scales

The use of large-scale maps or plans, without reference to taking off and calculating distances, should be familiar to the children. On such plans children can plot their journeys to and from school and identify streets and buildings. The work here is still informal, and recognizes that the children will begin to use maps fairly intensively at secondary levels. They can, however, begin to associate map measures with distances. This activity is marginally practicable in the junior school and could be omitted.

The work refers back to the simple use of scale as in Activity 45. Since a primary school cannot usually afford a set of maps for a class, the task is more suitable as an activity for a few children, working in pairs. Get the children to measure a route on the map in two ways, with a piece of thread and, if possible, a map wheel.

From this point onwards, the teacher needs to decide what to do according to the map actually available. The new 1:50 000 O.S. maps are now being issued for all districts, but most maps likely to be on hand in primary schools are in miles and inches. Point out to the children that actual measurement is not often needed on a map, since the scale printed on it is usually marked in kilometres or miles (or both). This is one of those situations in school where a brief demonstration of what is wanted is the most effective way of getting the children to start work. They can then practice finding distances using a piece of thread or a map wheel. The map wheel is normally used by starting at zero, running it along the route, and then running it backwards along a rule until it returns to zero again. This gives the distance in centimetres, although if it is run back along the map scale it will give the distance directly; this is the better method.

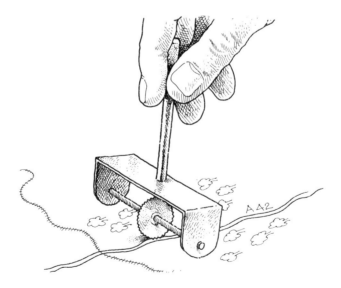

The work in distance should be extended by estimating the time to walk, cycle, or travel along given routes.

57 *Average height: measure in use*

This activity, useful when averages are being discussed, is most easily done with data taken from the personal measure books of a group of children.

Get pairs of children to prepare two strips of paper one-metre long; have one of them cut into five unequal pieces, each a whole number of centimetres. Have each piece marked with its length, so that the children can agree that the total length is 100 cm. Put end to end, the pieces would be as long as the uncut metre.

Now have this uncut metre cut into five equal 20 cm pieces. These have the same total length as the others, but are all equal. This length of 20 cm of the five equal pieces is called the *average length* of the five unequal pieces. The average value of any set of readings is what one would get if, coming to the same total, they were all equal.

1	2	3	4	5

five unequal lengths

1	2	3	4	5

five equal lengths

Now have a set of five strips which total any multiple of 5 cm other than a metre, and ask whether the average length could be found without cutting up another strip. We are aiming of course at totalling the lengths and dividing the result by five. This is a familiar arithmetical process, but the activity gives it a tangible setting.

Take further examples and finish by letting groups of four or five children find their average height, using data from their personal measure books. Take the result to the nearest centimetre, although one place of decimals could be used if understood. If so it would be a useful exercise in decimals although the overall accuracy is not likely to yield a reliable answer.

58   *Area and perimeter: how rectangles grow*

This is a useful and important exercise, but is very difficult to do practically. It is often described as an activity done with a loop of string; although pinning it out to form successive rectangles which can be measured in whole centimetres requires nimble fingers and considerable judgement. Good results can more easily be got by drawing the rectangles on squared paper, but this loses the concrete fixed perimeter provided by the string. The activity as described is an attempt at compromise.

Get the children to mark off on a sheet of graph paper pinned to a piece of soft wallboard a square 10 units by 10 units. We are going to work in units and squares so the actual size of the squares does not matter: 2 cm or one-inch squares are suitable. Now ask them to put four pins into the four corners and tie a loop of strong thread round them. As each question is asked and answered let the children shift the pins to match the new rectangles. Discuss the area and perimeter of the square, then ask:

If we make the square into a rectangle with one side only 8 units, what must the other sides be so that the perimeter stays the same?

Get the rectangle drawn up and repeat for rectangles 6 by 14, 4 by 16, and so on.

It is easier to work with a half-perimeter, tabulating the results neatly. The topic is very suitable for a class exercise led by the teacher, and serves to remind us that such activities still have their place in school.

Area of rectangle (half perimeter 20 units)		
LENGTH	BREADTH	AREA
10	10	100
12	8	96
14	6	84
16	4	
18	2	

From the results a graph can be drawn, plotting, for a rectangle of constant perimeter, the area against the length of one of the sides. For the results given the graph is shown by the solid line.

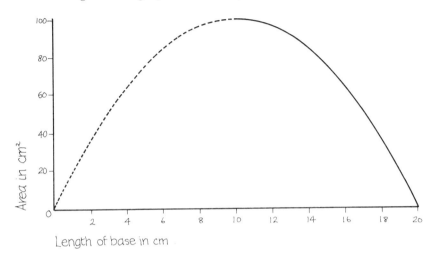

The graph at once suggests letting the original length decrease with a corresponding increase in breadth, which a child would then probably call height. Children could easily be confused by the use of words such as length, breadth, height, width, long, narrow, tall, thin. . . . They would tend to apply several of these words to various stages in forming the rectangles. What we are really concerned with is the length of the sides. The word 'base' might help, although here it suggests that the rectangle must be on a horizontal base line. Only controlled discussion can resolve the difficulty. The final graph is completed by the section shown dotted in the diagram.

The work can be extended in various ways. Possible questions could be:

Do we need to work out what would happen if we chose a side 5 units long, or could we get the answer from the graph?

Does it make sense to talk of one side zero? (What would happen if we stretched the loop out as a double thread?)

What shape gives the largest area?

This approach gives the opportunity to discuss interpolation and extrapolation of graphs, in this example taking the curve to zero and

20. Although the activity does not involve measurement as such, it leads on to later work on the ratios between linear dimensions and areas and volumes, a topic sometimes attempted at this stage but which is better left till more work has been done in practical science or craft, where it is an important principle.

59 *Three dimensions: the units of volume*

This can be thought of as a Stage Four activity deferred because it is more difficult to handle. It could be done in two parts, of which the first could be introduced earlier.

Use plastic centicubes or the one-centimetre cube units of sets of arithmetic blocks or rods. Build up blocks of about the size of match-boxes and discuss the number of unit cubes that make them up. Clearly the work extends that already done on area in whole centimetre squares, but we now have, in turn, the number of cubes in a row, the number of rows in a layer, and the number of layers in the block. The activity should not be attempted until the work in area has been thoroughly assimilated. Show the children two ways of setting out the work, say, for a 3 × 4 × 6 block

Number of cubes in a row	= 6
Number of rows in a layer	= 4
Number of layers	= 3
Total number of cubes	= 3 × 4 × 6
	= 72

Now relate the count of cubes to volume in centimetres cubed and write finally

length of block	=	6 cm
breadth of block	=	4 cm
depth of block	=	3 cm
volume of block	=	$(3 \times 4 \times 6)$ cm³
	=	72 cm³

It requires an effort to avoid the statement 'volume equals length times breadth times thickness', particularly since it gives the correct result in this example! Unfortunately the form of words often becomes associated with volume in general, and is applied to any shape such as a pebble. We are dealing with the oblong block or cuboid *only*, and this should be made clear at all stages.

Extend the work by considering very simple examples where dimensions are not whole number of centimetres, referring the pupils to the centimetre/millimetre graph paper of Activity 54. If the earlier activity work has been adequate, they should be able to visualize the set-up in millimetre cubes, and arrive at the correct volume for a block given as, say, 2.1 cm by 3.2 cm by 1.5 cm. The children should be able to suggest working in millimetres, so that the arithmetic reduces to finding the product of 21, 32, and 15, which is then the volume in millimetres cubed.

A matchbox-sized cuboid made by the teacher from cm/mm graph paper would suggest the arrangement of the stacked millimetre cubes. Arrange three one-metre dowel rods by fastening them into a block of wood having three drilled holes at right angles to one another. These show the dimensions of a metre cube: it is easy for the class to supply the other nine edges in their imaginations. Discuss with the class the number of centimetre cubes that go to make up the metre cube. If a 10 cm 'thousand cube' is available from number block apparatus, put one into the corner of the frame, so that the structure of 10 layers each with 10 rows of 10 'thousand cubes' can be visualized. The metre cubed will contain one thousand of these large cubes and hence one million of the centimetre cubes and one thousand million millimetre cubes. Instead of the 'thousand cube', a decimetre cube made from cm/mm graph paper put into the corner would suggest the stacking of the millimetre cubes.

The work with the growth of cubes does not lead to useful graphical work because it is not easy to get a suitable scale. The cubes of 1, 2, 3, 4, . . . grow so rapidly that the points soon run off the paper and are so far apart that it is not easy to draw a curve through them.

60 Volume and capacity: litres and decimetres

The litre is by definition a decimetre cubed. This activity is intended to underline the fact, not to present it as a 'discovery' of the relationship.

Cut or ask the children to cut some 10 cm lengths of milk straws. Since each is one-tenth of a metre, the length is called a decimetre and is sometimes used as a convenient unit (it is used for example for the draughts of ships so that we know what depth of water they need to enter a harbour). The length will also be quite familiar to children who use Cuisenaire or other rods.

Each child doing the activity should prepare 12 straws and join them with pieces of pipe cleaner (see *Children Learning Geometry*: Activity 39) to form a decimetre cube, which will then be the same size as the thousand block used in the last activity.

The children can be told the name of the unit, although they will not use it very often. If they imagine a trough the same size as their cubical frame, it would hold exactly one litre. It is probably worth while to have a commercially produced plastic decimetre cube that can be filled with water to demonstrate this. Since one thousand centimetre cubes make up the decimetre cube, the children should see why the millilitre is so named: one millilitre is equivalent to a centimetre cube.

61 Capacity and mass: a convenient relationship

This activity extends the previous work, but is given separately since it is better not done at the same time.

Not only is the litre, as a measure of capacity, related to the volume of solids occupying the same amount of space, but it is also related to the *mass* of the water it measures. Here again, it is a matter of telling the children. There is no reason why they should not weigh a litre of water, but they are most likely to 'discover' that it is 982 g or anything but 1 kg.

This could well be an opportunity to discuss accuracy and its limitations. It is easier to work the other way.

Using a compression balance with a pointer scale and the 1 dm cube plastic box, it is possible to show that 1 kg or 1000 g of water does in fact fill the box. Add the water until the scale reading increases by exactly 1000 g and show that the box is indeed full, then repeat with a thin plastic squash bottle of stated capacity 1 *l*. This cross-checks what has already been said about the litre and the decimetre cubed and avoids the inaccuracies of the first activity, since the fullness of the box is less critical. Now ask the series of questions:

If we fill this cubic decimetre with water, what do the contents weigh?

How many cubic decimetres fill up the metre cube?

How many kilograms of water would fill the metre cube?

Establish then that the metre cubed full of water weighs 1000 kg.

This is called the megagram (Mg) or more commonly the metric tonne. Question:

A school swimming pool is 1 m deep, 8 m long, and 3 m wide. How many tonnes of water does it hold?

At this point the word 'density' can be informally introduced. A substance is 'dense' if it is heavy for its size. Iron is denser than wood, and wood is denser than expanded plastic. If blocks of materials of the same size can be obtained, the children can handle them and accept the intuitive distinctions. The qualitative aspects can then be discussed.

The density of water is 1 g per millilitre or 1000 kg per metre cubed. Give a few other examples in passing: iron is 7860 kg per metre cubed, so that a block of iron that fitted the dowel frame of Activity 56 would weigh nearly as much as 8 tonnes of coal. Ice, on the other hand, is lighter than water, and a block of this size would weigh 920 kg, not 1000 kg. This is why ice floats. The topic can be followed beyond this point if teacher and class are interested, but it is one that will be picked up again formally at secondary level.

62   *Timetable time: the 24-hour clock*

This is one more activity that needs to be carried on as required over the school year, until the 24-hour notation becomes so familiar that it is read without effort.

Collect a few old bus, airline, or railway timetables or excursion handbills, and cut out suitable sections to stick on cards. A double folder is best. The timetable section can be stuck on the left inside, a sheet of questions on the right. Examples of questions:

Say when the first bus leaves Woodrow Street Bus Station on Sunday morning.

When does it arrive at Renton?

How long does the journey take?

If you miss this bus, how long must you wait for the next?

Convert all the times for the Sunday service to Shenton to the ordinary 12-hour clock.

The questions should, where possible, be answered in complete sentences, and the exercises should call for good writing, spelling, and neat layout. This is often a major problem! To persuade a pupil to write 'The bus leaves Woodrow Street at 10 25' instead of scribbling '10 25' is a continuing struggle. Possibly it need not be done all the time, but some at least of the practice should be an exercise in presenting information. This is often a major problem! To persuade a pupil to write mation.

Note that many children have difficulty in calculating time intervals. If a bus leaves at 09 41 and arrives at 11 20, how long does the journey take? At one time this was taught by one of the standard methods of subtraction using a mixed base, and children would have been expected to set down something like this or its equivalent in another algorithm:

$$\begin{array}{r} 60 \\ 0 \quad 1 \\ 1\not{1}.\not{2}\not{0} \quad 10 \\ 9.41 \\ \hline 1.39 \\ \hline \end{array}$$

We strongly recommend that such calculations should be done from the start using mental arithmetic. No bus enquiry clerk will produce a sheet of paper and use the method of decomposition! It takes longer to become adept and there is less written work to show, but the skill will at least be relevant to the use of timetables. For example, one can say '9 41 up to 10 00 is 19 minutes, so on to 11 20 is 1 hour 20 plus 19, which is 1 hour 39 minutes in all.' Pupils can be given oral sessions of this kind of work, for which they can be asked to write down numerical answers only, without working. Examples need to be carefully graded. The one given is difficult.

63   *How hot is it?: using a thermometer*

Activity 29 has already introduced the thermometer and we hope that children will by now be quite familiar with reading it. This activity encourages children to use thermometers, introducing them to an instrument of scientific measurement.

One can begin by referring to the wall thermometer, producing a laboratory version for inspection by the children.

We now discuss the thermometer as an accurate instrument. It is, of course, a fragile one and needs to be carefully handled. Help a capable pair of children to rig up a thermometer in a can of very hot water, and get them to take a reading every few minutes, making a neat table of the results. Afterwards discuss putting these results on a graph, with time in minutes on the horizontal and temperature on the vertical axis.

Note that the cooling will be due to three separate factors, to radiation, convection currents in the liquid and in the air, and evaporation from the surface, so that it is not an experiment that can be considered formally. It is best regarded merely as a familiarization exercise that will give the children a start when they begin their secondary courses. A stirring device is not needed for our purpose.

During winter, bring in some snow in a couple of jars. Get a child to put a thermometer in one and note that the temperature is indeed zero or just below. Mix salt with the other jar and stir (not with the thermometer!) until the contents just become liquid or at least slushy, then ask the children what the temperature is likely to be. They will probably suggest just above zero, since the salt has made the ice melt, and they will be intrigued to note that it is now well below zero. The explanation, as reference to a textbook of physics will show, is not easy to put across even to older pupils, but the reading of the thermometer is there for all to see.

Try to get the children to keep in mind a few key temperatures to act as reference points in judging whether things are hot or cold. All other work in temperature, formal measurement as apart from incidental use in discussing melting, climate, and so on, is better left to secondary stages.

64 *Difference in direction: measuring angles*

The measurement of angles in degrees, like the measure of lengths in millimetres, causes difficulty to young children because of the smallness of the unit. On an ordinary semicircular school protractor the degree intervals are about the width of pencil lines as most children draw them.

Refer to Activities 25, 34, and 46 in *Children Learning Geometry* for a nonquantitative approach to angle and the use of compass points.

Make an angle indicator by pivoting two pieces of lath with a thumbscrew to tighten the joint. Use it to discuss acute and obtuse angles and the right angle, relating the work to the points of the compass. The classroom approach to compass points is fully discussed in *Children Learning Geometry*.

Introduce the degree as 1/360 part of a complete turn. Let children inspect 360° degree circular protractors, the larger the better, so that they can see how they are numbered. Issue sheets with a number of duplicated lines at various angles, and discuss with small groups of children how to use the circular protractor to find the angles between the lines.

Follow on by asking the children to draw lines at given angles to a base line. The work at this stage could include reflex angles greater than 180°; the children must then learn to mark the angle with an arc to show which part is wanted.

Extend the work by discussing the modern 000° to 359° three-digit bearing system. This has entirely replaced the older numerical systems in actual practice, and there is no need to mention them. Let the children draw up an imaginary island or coast with ample rocks, wrecks, look-out points, and buried treasure, by choosing a suitable scale. They can take off and list the bearings and distances of features from look-out and landing points.

The children should draw in a N–S line through each point from which a bearing is taken, drawing a second line through the rock or wreck, and should use kilometres for distances. In practice, on a chart, one would take distances in nautical miles, which for technical reasons remain in use in navigation. Few children will have handled or even seen a nautical chart, and it is a good idea to have one on display. Charts are constantly being corrected and amended, and one can usually obtain obsolete sheets very cheaply from chart agents.

Note that we do not recommend drawing up triangles or quadrilaterals whose angles are to be measured and added together. Cumulative errors lead to inconsistent results, and the attempt misses altogether the deductive nature of the well-known statement about the angle sums of polygons in geometry. Primary children are not yet ready for this, although they can use cut-out triangles and tear off the corners or fold them inwards, putting them together to get a straight angle.

Children using measure: a note on primary science

If properly planned and spread over the primary years the five stages of activities should make a child feel at home in any task calling for measurement that is likely to be met, taking into account age, general level of education, and the needs of daily life. Although they provide a lot of incidental practice, our activities are in the main intended to introduce measures. They must be kept in use, linked up with the rest of mathematics, kept in line with computational skills, and applied to those situations that call for them.

It is only when children can use measure incidentally in performing tasks that require it that they can be thought of as operationally competent.

Primary science offers a clear field for the use of measure. In science, a child learns to make the transition from qualitative observations to quantitative descriptions. A child readily sees that a long pendulum beats more slowly than a short one, but in making a formal study of the pendulum, measuring its length and timing its swings, becomes gradually aware of one aspect of the scientific method, of the possibility of reducing conclusions to numerical statements.

There should be no difficulty in linking science and mathematics in a primary school, since a teacher who is responsible for the entire work of a class can integrate these studies and use one to help forward the others. This book, however, is not a work on primary science and the following activities are not developed but only suggested.

They are chosen from a possible systematic programme for school science because their value for scientific education is only arrived at through measure and its use in computational or graphical mathematics. For this reason they have a bias towards mechanics, but we have tried to introduce work in other branches where measure can widen the scope of the studies.

It is unfortunately true that not all schools have a programme of scientific studies. It is not indeed easy to devise one at primary level, where the children have to make the transition from the 'nature table' to the technically equipped laboratory with its 'set piece' experiments. We take it, though, that most of the following activities would be included in any programme and given suitable treatment. They are meant to be selected by the teacher to underline some aspect of measure or computation with measure, and are not intended as a sort of buffet table from which children can choose what they would like. A few of the activities may not be thought of as particularly 'scientific' but all of them require children to apply the skills of measurement.

Absorption of liquid

Weigh 500 g of dry sand into a can with small holes in the bottom. Flood with plenty of water and allow to drain through the holes till no more water runs out. Weigh again, and express water absorbed as percentage of the dry weight and also of the total weight. Repeat for pieces of dry cloth, porous objects, samples of small coal, and so on.

Food intake

Weigh daily the solid food given to the class guinea pigs or gerbils and calculate their weekly and monthly intakes. Weigh the animals at regular intervals and construct a graph of body weight.

Length of pendulum

Using a stop-clock, plot the time in seconds for 20 swings for pendulums of various lengths, and construct a graph with the length on the horizontal axis (it is usual to choose the horizontal axis for the quantity whose change is under control: in this case the length, which is adjusted before retiming).

Children are often asked to investigate whether the arc of swing or the mass of the bob make any difference to the results. It is better to let them do this as a rough experiment with two pendulums of the same length. If the arc is large there is a marked discrepancy over 20 swings, and it requires very careful attention to detail to alter the mass of the bob without changing the effective length of the pendulum. If, however, two pendulums are swung together a few times, one can see that neither the arc of swing nor the mass is making an effective difference.

Compound pendulum

Drill small holes at 5 cm intervals along the centre line of a metre rod. Insert lengths of steel knitting needle through the end hole and, with the pin supported on hard edges if possible, time the rule for 20 swings. Shorten the pendulum by moving down to the next hole, and note the increased frequency of swing. Continue the experiment and plot the results, time of 20 swings against distance of pivot from the end of rule. This is usually an interesting activity for the children, since it gives a result they do not at first expect.

Quadrat studies

On the school field or any other accessible plot, measure out a rectangular area with corner posts using metric tapes. The area should be as large as conveniently possible. Throw quadrats (see Activity 52) and count the population of a chosen plant. (Choose a plant, after preliminary inspection, that will give suitable results: this is a training exercise not an agricultural survey!) Calculate the measured average incidence of the plant per metre squared and estimate the total population of the area. A half-metre quadrat may be easier to handle, and provides one more simple stage in the calculation.

Shadow sticks

Erect a pointed shadow stick in a sunny spot and measure the length of the shadow at intervals. Plot the length against time. Erect the stick at the centre of the side of a large sheet of paper fastened to the ground. Mark the direction of the shadow at intervals. Later, measure the angles made with the reference side and plot against time. Use the marked paper to make a sundial by drawing in the shadow lines carefully and marking the times clearly. Set the paper and stick up the next day and check that one can tell the time from it.

Height/shadow ratios

Have poles of length 1, 1½, 2, 2½ metres. Have pairs of children measure their lengths and then, working together at the same time, their shadows. Use simple plumb lines to check that they are held upright. Plot height against shadow length and discuss possible conclusions.

Volume by displacement

Use a displacement can (a can with a downward sloping spout as in the diagram) to find the volume of large beach pebbles. Avoid porous objects or those that float on water: these need experimental techniques that go beyond our intentions. If the pebbles are then weighed, one can discuss density as introduced in Activity 61, although the masses and volumes as actually recorded are likely to give numerically awkward values if used for computation.

Growth of cubes

Where multibase arithmetic blocks are available, sort out samples of cubes from base 10 down to base 2. Do not use the unit cubes from the set since they are too light for accurate results. Plot weight against side of cube. Larger cubes can be made by using eight smaller ones. Plastic cubes of mass 1 g and volume 1 cm² are now available.

Reaction times

Form a large group into a ring holding hands. On the word go some-one starts a stop-clock and squeezes the hand of the next child, who passes on the squeeze. When the squeeze has returned to the first squeezer the clock is stopped. Repeat the experiment, trying to improve the time, and plot the results as trial number against seconds delay.

The children usually react vigorously when passing on the squeeze; so that by standing back one can usually *see* the squeeze travelling and keep an eye open for discontinuities. The graph will normally show an improved performance which evens out and then slows up as the effort leads to reaction fatigue.

Tree rings

If a tree is being felled locally those doing the job will often slice off a section of the trunk if asked by a school. Count the tree rings and discuss the dates that arise. It might be necessary to smooth the sawn section a little with a Surform tool to bring out the rings clearly.

Suction

Have a length of straight plastic tube dipping into a bucket of water and let a child suck up the water as far as possible. Let another child match up this height with a plumb line and measure it in centimetres. Repeat when standing further from the bucket so that the tube is more slanted. Compare vertical heights. Children can record the heights in their personal measures book.

Glider competition

The children can each design and make a glider from one sheet of A4 paper provided, using paper clips *ad lib* for balance weights if needed. On a windless day (or indoors in the hall) have them release their gliders from a marked spot and measure with a tape the distance flown. Get each competitor to write a specification of the glider as trimmed for flight: total weight, length, wing span.

Heights of trees

Investigate the heights of trees using a clinometer. There are several plastic models on the market. Discuss the two techniques: stepping back until the angle of elevation is 45° and then measuring the distance to the base of the trunk, or else measuring the angle for any convenient distance and making a scale diagram with ruler and protractor. The children will need to draw a diagram to see that when the angle is 45° the height of the tree above eye level is the same as the distance from its base. If they use the second method, they may need a class or group demonstration of working from a scale diagram using ruler and pro-tractor.

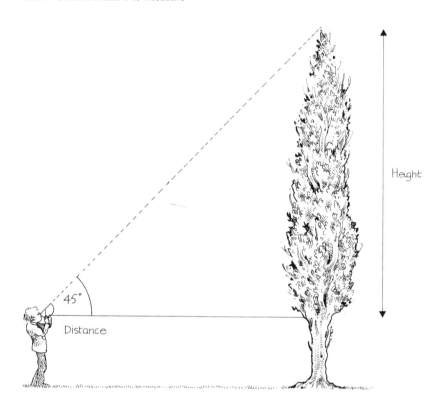

Compare the method of using a vertical shadow stick of known height and measuring the shadow of the tree. The ratio of height to shadow for the stick is the same as that of the unknown height and measured shadow of the tree. Note that the second and third methods provide practice in different kinds of measure and computation and that the first requires no computation at all.

Conclusion

If children can tackle these or similar activities, performing the required measurements with sufficient precision and the required calculations with sufficient accuracy, there should be no great difficulty in undertaking secondary studies in laboratories or workshops.

The real problem seems to be not that of getting the children to do the measures in the first place but of keeping their skills alive and practised. Children of primary age, moved from one country to another, can forget the correct use of their own native language if they stop using it, and we must not expect more of a skill taught in a classroom. Fortunately, the art of measurement is a general skill which easily adapts itself to any specific need, and as long as the primary school can keep measures in use at the levels we have discussed there should be no problems in extending the work to secondary stages. Such problems as arose would have a deeper source than lack of background in the fundamental skills.

Appendix 1

Appendix 2

The activities listed by topics

This list classifies the activities according to their main topic. The later activities may draw incidentally on several measures already worked at; but in general they are devised for a specific stage in learning to use one measure.

Length and distance

1, 2, 3, 4, 5, 6, 16, 17, 22, 27, 28, 31, 32, 33, 34, 43, 44, 45, 47, 53, 55, 56, 57, 58.

Capacity

7, 8, 26, 30, 42, 61.

Area

10, 19, 20, 21, 22, 24, 37, 38, 48, 49, 52, 54, 58.

Volume

7, 23, 24, 59, 60

Mass and weight

9, 15, 18, 35, 36, 46, 61

Time

11, 12, 13, 14, 25, 39, 40, 41, 50, 62.

Temperature

29, 63.

Angle

64.

General

51.

Appendix 3

Apparatus in the classroom

Although we are all agreed that the primary classroom should have a wide selection of apparatus and materials, there are still two schools of thought about what it should be. On the one hand we have the extemporizers and junk collectors, on the other those who want the children to have access to properly designed purpose-made equipment that will introduce them to professional standards of construction and use. The first collect cans, bottles, ends of wallpaper, surplus floortiles, old clocks, cartons, and discarded kitchen gear; the other buy clinometers, geared demonstration clocks, metric tapes, and collections of plastic shapes.

It seems clear that we need both; but in these times of shortage of funds we should keep expenditure down to the essential items. A large plastic demonstration micrometer, for example, is a good thing to have if one can afford it, but educationally it is worth far less than its value in marker pens and assorted papers. Accurate tape measures, however, cannot be extemporized, but the children do need them. They must be brought in as purchased items.

All schools and teachers' centres have, or can have, access to the very comprehensive catalogues sent out by education suppliers. These contain far more in the way of specially designed apparatus than any one school could want or afford. Most schools, at least in urban areas, could equip themselves with useful measures more cheaply if there were less inflexible arrangements for petty cash purchases in local stores. Browsing through the catalogues often suggests apparatus that can be got together in this way.

The names of the major suppliers must be the staff room equivalent of 'household names', and we shall not give them, but we add here a list of items that one would like to find in the primary classroom at some stage in a child's school career.

metre rods, divided decimetres

metre rods, divided centimetres

centimetre rules, alternate intervals coloured

centimetre/millimetre rules (not for general use)

metric tapes

rope marked out in metres

string, cords, and tapes of all kinds

height measurer

foot length measurer

trundle wheel (one metre)

maps, large scale

area grids in plastic

tessellation shapes in plastic and coloured paper

calibrated jugs

compression balance

equal arm balance

masses 10 g plastic, samples of others up to 1 kg

letter type balances

open plastic box, 1 dm³ internal

protractors, plastic 360°

clinometer

mounted globe

clock face stamp

geared clock (or old wall clock)

sand timer

'pinger' kitchen timer

stop-clock, large face

dry goods of all kinds – peas, beans, conkers, acorns, rice, sand, counters, and shells

parcels and objects for weighing

tiles

matchboxes

used stamps

cartons and boxes of all kinds

bottles and jars in m*l* or c*l* sizes

squared paper, all sizes (cm/mm paper for occasional use only)

plasticene

It is tempting to add 'etc.', and leave it at that! A lot that finds its way into refuse bins could at least find a temporary use in a classroom. Examples would be the squared backing paper of flexible plastic coatings, the ingeniously folded cartons used for sweets or cosmetics, or smoothly opened cans from the school kitchen. The list is, of course, to be extended by all the items useful in topics other than measure.

Bibliography

These books provide useful background information on the history of measurement. Teachers might consider a few of them for class or school libraries. The last two books are essential references for a modern syllabus.

Bowman, M.E., *Romance in arithmetic: a history of weights, measures and calculation,* University of London Press, 1969.

Hogben, L., *Man must measure,* Rathbone Books, 1955.

Hood, P., *How time is measured,* Oxford, 1955.

Smith, T., *The story of measurement,* Basil Blackwell First Series, 1955; Second Series, 1959.

Srivastava, J.J., *Weighing and measuring,* A & C Black, 1971.

Nuffield Mathematics Project, published by W. and R. Chambers and John Murray,

𝒱   Beginnings, 1967.

𝒱   Shape and size, 1967.

𝒱   Shape and size, 1968.

𝒱   Shape and size, 1971.

Metrication Board, *How to write metric,* HMSO, 1977.

The Royal Society, *Metric units in primary schools,* 1969.

Index

This index does not contain, except for purposes of cross-reference, entries for items which are more conveniently found by referring to Appendices 1 and 2.

The Harper Education Series has been designed to meet the needs of students following initial courses in teacher education at colleges and in University departments of education, as well as the interests of practising teachers.

All volumes in the series are based firmly in the practice of education and deal, in a miltidisciplinary way, with practical classroom issues, school organisation and aspects of the curriculum.

Topics in the series are wide ranging, as the list of current titles indicates. In all cases the authors have set out to discuss current educational developments and show how practice is changing in the light of recent research and educational thinking. Theoretical discussions, supported by an examination of recent research and literature in the relevant fields, arise out of a consideration of classroom practice.

Care is taken to present specialist topics to the non-specialist reader in a style that is lucid and presentable. Extensive bibliographies are suppled to enable readers to pursue any given topic further.

Meriel Downey, General Editor

Teachers of Mathematics: Some Aspects of Professional Life
edited by Hilary Shuard, Homerton College, Cambridge, and
Douglas Quadling, Cambridge Institute of Education

Clever Children in Comprehensive Schools by Auriol Stevens,
Education Correspondent, The Observer

Values and Evaluation in Education edited by R. Straughan
and J. Wrigley,University of Reading

Middle Schools: Origins, Ideology and Practice edited by L.
Tickle and A. Hargreaves, Middle Schools Research Group

The Mathematical Education Trust

Mathematics Teaching: Theory in Practice by T.H.F.
Brissended, University College of Swansea

**Children Learning Geometry: Foundation Activities in Shape,
A Handbook for Teachers** edited by J.A. Glenn

Teaching Primary Mathematics: Strategy and Evaluation
edited by J.A. Glenn

The Third R: Towards a Numerate Society edited by J.A. Glenn

Teachers of Mathematics: Some Aspects of Professional Life
edited by Hilary Shuard, Homerton College, Cambridge, and
Douglas Quadling, Cambridge Institute of Education